INFORMATION THEORY AND
THE LIVING SYSTEM

INFORMATION THEORY
AND THE LIVING SYSTEM

LILA L. GATLIN

COLUMBIA UNIVERSITY PRESS · 1972

NEW YORK AND LONDON

Lila L. Gatlin is Assistant Research Biophysicist
at the Space Sciences Laboratory, the University
of California at Berkeley

Copyright © 1972 Columbia University Press

Library of Congress Cataloging in Publication Data
Gatlin, Lila L.   1928–
    Information theory and the living system.
    Bibliography: p. 205
    1. Information theory in biology. I. Title.
QH507.G37   574.1   76–187030   ISBN 0–231–03634–5
Printed in the United States of America

*AMS 1970 Subject Classifications 60G10, 60J10
90D45, 92–02, 92A05, 94–02, and 94A15*

To Carl, Amy, Jeff,
Laura, Jennifer

PREFACE

I began this book as a series of lectures to graduate students in biology at Bryn Mawr in the fall of 1967. I completed it at the Space Sciences Laboratory in Berkeley, largely as the result of the encouragement of Dr. Thomas H. Jukes, and submitted it for review in the summer of 1970. Review, reappraisal, and rewriting bring us near the summer of 1971. I am fully aware of the ways in which the work is unfinished, but then it will always be thus. I am also aware of the ways in which it is finished. For example, D_1 and D_2 are invariant functional forms. As I continued to work on the book, so many new possibilities for future work presented themselves that, if I do not publish it now, it will be more "unfinished" next year than it is now.

Also, in writing books one can get too bogged down in detailing the past at the expense of curtailing one's future work. Books, like living organisms, should be constantly evolving, and their time of publication should optimize the variables mentioned above.

I can no longer resist the lure of future work, and I must be unencumbered by the responsibility of a book. Let us hope that by publishing at this time I have succeeded in minimizing the maximum error and maximizing the minimum creativity of this work.

I wish to thank Thomas H. Jukes, Hans J. Bremermann, David E. Kohne, Robert A. Elton, Thomas A. Reichert, Richard E. Holmquist, and Temple F. Smith for reading and commenting on various sections of the book. I also wish to thank my reviewers, whose comments are responsible for major revisions in the book, hopefully for the better.

Summer, 1971   *Lila L. Gatlin*

CONTENTS

INFORMATION THEORY AND THE LIVING SYSTEM

1

THE INFORMATION PROCESSING
SYSTEM

Life may be defined operationally as an information processing system—a structural hierarchy of functioning units—that has acquired through evolution the ability to store and process the *information* necessary for its own accurate reproduction. The key word in the definition is *information*. This definition, like all definitions of life, is relative to the environment. My reference system is the natural environment we find on this planet. However, I do not think that life has ever been defined even operationally in terms of information. This entire book constitutes a first step toward such a definition.

It is obvious from a simple inspection of the language of the biologist that the word information is indispensable. It appears on page after page of any modern biology textbook. For example, Stent (1963) states that mutations are: "a sudden change in *informational content* of the hereditary substance"; "bacterial viruses are the carriers of a complex hereditary apparatus whose *informational* content must be very high"; "these viruses bring into the host cell sufficient genetic *information* for the construction of all enzymes . . ."; etc. The word is vital to the vocabulary of the biologist.

WHERE THE INFORMATION IS STORED

Today we know something about the mechanics of this information storage and processing. We know where the information is stored:

in the base sequence of DNA, or RNA in the case of some RNA viruses.

DNA, deoxyribonucleic acid, is a tremendous polymer with molecular weights up to a billion found within all living cells. It consists of two polynucleotide chains wound helically about a long central axis. The bases on opposite chains are joined through hydrogen bonds according to the constraint that adenine (A) bonds only to thymine (T) and cytosine (C) bonds only to guanine (G) (the Watson-Crick base-pairing rules). There are two hydrogen bonds between A and T and three hydrogen bonds between C and G. All other bonds in the molecule are covalent. There are no restrictions on the *sequence* of bases along the helix axis. The outer part of each strand, which is called the *backbone*, consists of a regular sequence of alternating deoxyribose-phosphate residues joined in phosphodiester linkage from the 3′ carbon atom of one deoxyribose ring to the 5′ carbon of the next. This pattern, shown in Fig. 1, endows the DNA strands with direction or polarity defined by the 3′ → 5′ direction of the deoxyribose-phosphate linkage. The two DNA strands have opposite polarity. The polarity of the DNA strands results in a most important mathematical property. When the statistical properties of the sequence of bases along a single strand are observed, as for example in measuring the frequencies of doublet or two-base sequences, these numbers have both magnitude and direction. Hence they are *vectors* not *scalars*.

A simplified representation of DNA structure is shown below.

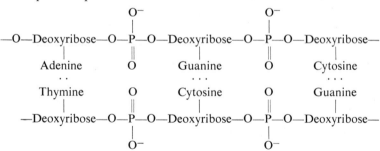

This can be abbreviated to

    —A—G—C—

     ·   ·   ·

    —T—C—G—

FIGURE 1

SCHEMATIC DNA STRUCTURE.

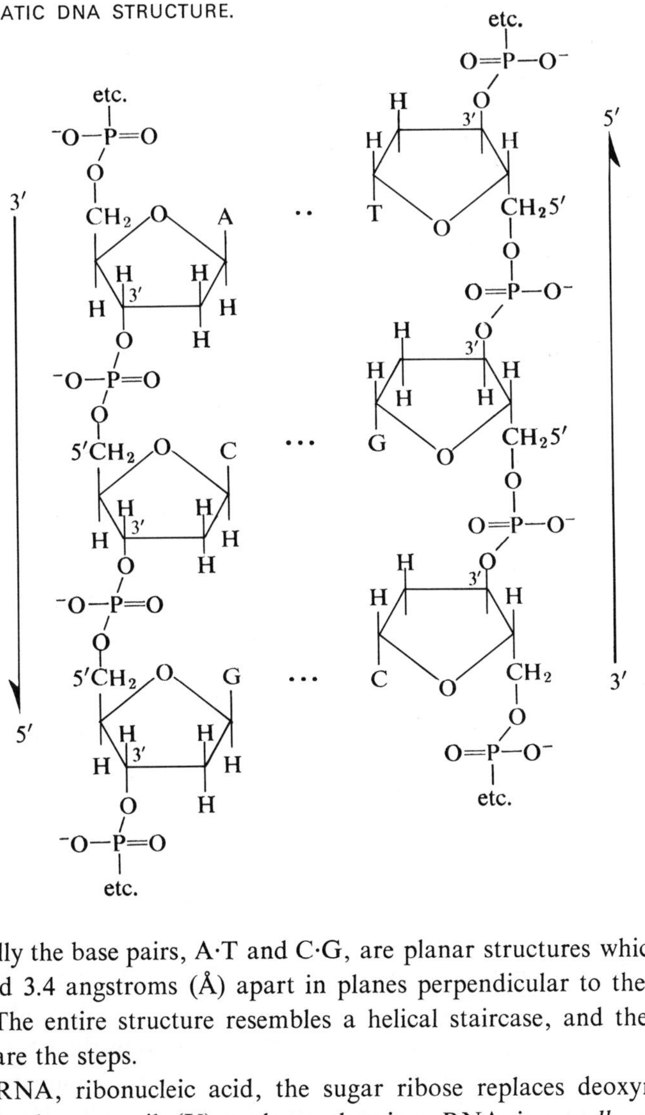

Actually the base pairs, A·T and C·G, are planar structures which are stacked 3.4 angstroms (Å) apart in planes perpendicular to the helix axis. The entire structure resembles a helical staircase, and the base pairs are the steps.

In RNA, ribonucleic acid, the sugar ribose replaces deoxyribose and the base uracil (U) replaces thymine. RNA is *usually* single-stranded but has a marked tendency to form double strands by looping back on itself, whereas DNA is usually double-stranded, although exceptions exist in both cases.

The two strands of DNA unwind during replication, and each strand acts as a template for the formation of a complementary strand according to the base-pairing rules. Two double strands are thus formed from one: each new double strand with an original "parent" strand and a new "daughter" strand. This process is mediated by enzymes and may be far more complicated than the simple Watson-Crick base-pairing rules would suggest. It does explain, however, how the base sequence of the DNA molecule could be exactly duplicated and passed down from parent to progeny. This is a necessary condition for the storage of genetic information.

Every individual living system has its own unique sequence of bases along its DNA chain, which we may regard as a sequence of symbols that stores information in the same manner as a sequence of letters in any language. In fact, one could regard language, an ordered sequence of symbols with a definite meaning, as the basis of all life, and the base sequence of DNA as a message in the genetic language.

The DNA alphabet contains only four letters, A, T, C, and G, but this does not limit the storage of information qualitatively since even a sequence of only two kinds of symbols from a binary alphabet can store any message in any language. The only limitation is that, the smaller the alphabet, the longer is the sequence necessary to transmit a given message.

DNA sequences are very long. The minimum DNA content per haploid cell (the genome size) ranges from about 10^4 base pairs for bacteria to over 10^9 base pairs for mammals (Britten and Davidson, 1969).

Since there are four kinds of DNA bases, over 4^{10^9} base sequences are possible for present-day organisms. This number is greater than the estimated number of particles in the universe.

The sheer magnitude of the DNA molecule should serve to warn us that a completely different approach may be necessary in studying its sequence than is necessary, for example, in studying the amino acid sequence of proteins, which seldom exceeds a few hundred amino acids and then only as aggregates of identical subunits. Hoyer and Roberts (1967) calculate that, even if the base sequence of DNA could be chemically determined at the rate of one base per second, it would take over 100 years to sequence one mammalian DNA. Furthermore, if we wished to compare two or more such sequences, we could not possibly do this "by hand." Yet we have no computers which even approach the necessary capacity to store and manipulate such data.

Thus it appears unlikely at the present time that we will ever study the properties of the total DNA of living systems, particularly of higher organisms, by simply determining their base sequences. Therefore it is particularly significant that in information theory we have a mathematical tool which can give us an insight into the informational properties of the base sequence as a whole of any large DNA molecule. In fact, the larger the molecule, hence the longer the sequence, the more accurately does classical theory apply. The experimental techniques necessary to obtain the statistical data are well within our grasp.

Several independent lines of evidence lead to the conclusion that DNA stores the hereditary information. The DNA of certain bacteria has the ability to "transform" other bacteria. If the DNA of one strain is purified and injected into another strain of the same species, the recipient bacterial strain displays a permanent hereditary alteration. This transforming activity is not affected by enzymes which degrade RNA or proteins but is readily destroyed by enzymes which degrade DNA. The amount of DNA is constant per diploid set of chromosomes for any eukaryotic organism (i.e., one with a nuclear membrane) and equal to twice that of the haploid germ cell. Bacteriophages, bacterial viruses, leave their protein mantle outside the bacterial cell they invade. This protein then can be removed from the system without changing the infectious process in the least. Clearly, it is the *information* stored in the small DNA molecule which is injected that literally takes control of and redirects the entire metabolic apparatus of the bacterial cell. The massive machine is submissive to the power of the controlling information.

The fact that DNA is a linear sequence of symbols and exactly duplicates this sequence each time the molecule reproduces itself confirms its role as the site of information storage. Finally, the growing body of experimental knowledge describing how the base sequence of DNA is transcribed and translated into the amino acid sequence of proteins essentially proves that the linear sequence of symbols in DNA stores hereditary information.

Although there is little doubt that DNA is the basic unit for the storage of genetic information, it is not yet known how the information acquired by the organism through learning is stored in the brain. Some workers believe that the basic unit of information storage is a linear sequence of symbols, as it is for genetic information. This basic unit is believed to be an RNA molecule or protein molecule synthesized in the brain during learning and is, in fact, transferable from one individual to another. Others believe that the basic memory unit of the brain is a three-dimensional structure such as a neural pathway. Perhaps both play a role, perhaps not. There is no general agreement as yet, and which alternative one prefers is a matter of faith. However, when we state that the genetic information which controls the synthesis of proteins and is passed down from parent to progeny is stored in the base sequence of DNA, this is not a matter of faith. It is an established scientific fact.

HOW THE INFORMATION IS TRANSMITTED

I have defined life as a system that both stores and processes the information necessary for its own reproduction. Under this definition a virus would not be regarded as alive since it can only store but cannot process information. It must enter a living cell to do so. Within the living cell is an intricate information processing machine.

Experimental efforts today are intent on delineating the inner workings of this machine, and we know a great deal about the mechanical details of how the information is transmitted, or translated into terms for protein synthesis. First the information stored in the base sequence of a DNA strand is copied by the formation of a strand of "messenger" RNA (mRNA) utilizing the single DNA strand as a template. The copying process, called *transcription*, is analogous to the DNA replication process. The base-pairing rules apply except that adenine on the DNA strand pairs with uracil on the RNA strand. As in replication, the two strands have opposite polarity. This process is carried out by a special enzyme, *RNA polymerase*, which attaches itself to a *transcription initiation site* on one strand of a DNA molecule. This site is identified by a special type of base sequence.

The enzyme travels along the DNA strand, producing a single strand of RNA as it proceeds, until it reaches a second special sequence of bases which carries information that instructs the transcribing operation to halt. The RNA molecule is then set free.

A double strand of DNA hence gives rise to a single strand of RNA as shown below:

DNA —A—C—T—T—A—C—C—G—A—A— $3' \rightarrow 5'$

.   .   .   .

—T—G—A—A—T—G—G—C—T—T— $5' \leftarrow 3'$

RNA —A—C—U—U—A—C—C—G—A—A $3' \rightarrow 5'$

The information is now in "working" form. The transcription process could be regarded as analogous to the process of retrieving information from the memory unit of a computer and copying it into other cells where it can be manipulated and possibly modified without damaging the original information.

The input to this information processing system is the base sequence of DNA, and the output is the amino acid sequence of protein. What transpires in between is complex, but it is clear that a major component of the translation mechanics is the "transfer" RNA (tRNA) molecule. Like mRNA, tRNA molecules are formed on the DNA template but, unlike mRNA, many of the bases are subsequently modified by methylation and other chemical changes. tRNA molecules are about 80 bases long and fold back to reassociate partially with themselves through Watson-Crick base pairing such that the planar projection of the structure resembles a clover leaf with base-paired arms which open into single-stranded loops. A typical tRNA molecule is shown in Fig. 2. On the central loop is a sequence of three bases called the anti-codon, which is complementary under modified base-pairing rules to a sequence of three bases on mRNA, called a codon, which specifies a particular amino acid.

For each of the 20 "coded" amino acids there is at least one specific enzyme, an amino acyl-tRNA synthetase, which "activates" the amino acid by attaching it to its specific tRNA molecule. It has been decided recently to rename this enzyme amino acyl-tRNA ligase. The enzyme must recognize the amino acid and at least some part of the base sequence of the tRNA molecule. The tRNA recognition site could not be just the anticodon. A common sequence, CCA, is added enzymatically to

FIGURE 2

SCHEMATIC tRNA MOLECULE.

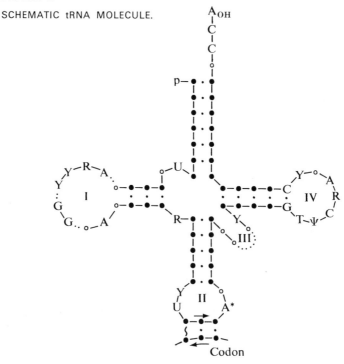

Codon

the 3' hydroxyl end of all tRNA molecules. The tRNA and amino acid are joined in ester linkage between the carboxyl group of the amino acid and the 2' or 3' hydroxyl group of this terminal adenylate.

The information directing the sequential assembly of amino acids in the protein chain resides in the mRNA molecule, which now becomes attached, presumably by hydrogen bonding, to complex structures called ribosomes. Since the mRNA is much longer than the ribosomal particles, it is usually "read" simultaneously at various points along its length by several ribosomes. This complex is called a *polysome.* The ribosomes are composed of ribosomal RNA and highly specific proteins. There are probably two sites on the ribosome that can be occupied by tRNA molecules, the donor and acceptor sites shown as X and Y, respectively, in Fig. 3. The donor site is first occupied by the codon AUG on the mRNA and a special tRNA species. In *Escherichia coli* this is formylmethionine tRNA, which carries the amino acid

FIGURE 3

PROTEIN SYNTHESIS. X = DONOR SITE, Y = ACCEPTOR SITE.

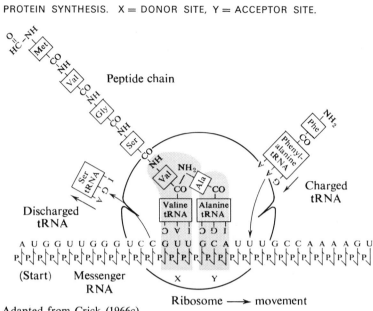

Adapted from Crick (1966c).

methionine with a formyl (—CHO) group attached. This process initiates the growth of a polypeptide chain and this first methionine is often removed before or soon after the protein molecule is complete. The special "initiation" tRNA molecules used by eukaryotic organisms carry methionine that is not formylated. Charged amino acyl-tRNA molecules then "move in" to the acceptor site and their anticodon base pairs by hydrogen bonding with the codon on mRNA which, according to the genetic code, corresponds to the amino acid esterified to the tRNA molecule. Thus there is no actual physical association during protein synthesis between the codon on mRNA and the amino acid it specifies in protein. This tRNA molecule is sometimes called an "adaptor," as originally proposed by Crick (1955). The assembly of amino acids then takes place sequentially as the carboxyl group of the donor amino acid is transferred from its ester linkage in tRNA to peptide linkage with the amino group of the acceptor amino acid. The donor amino acid is thus released from its specific tRNA. The complex now shifts by one frame of three bases, and the previous acceptor amino acid becomes the new donor and a new amino acyl-tRNA species moves into the next codon frame. This entire process is diagrammed in Fig. 3. The process continues until a chain termination sequence on the mRNA containing at least one of the codons UAA, UAG, or UGA is reached. The complex is then released from the ribosome.

This entire *translation* process requires GTP (guanosine triphosphate) and a number of soluble protein "factors" which mediate initiation, translocation, termination, and perhaps even specificity of codon recognition. New protein "factors" are still being described. Hence we do not yet know the extent of the complexity of the translation process.

Complexities exist in the process of recognition between codon and anticodon. It is now known that many tRNA species can "recognize" two or more codons which differ from each other by the base in the third position of the codon. This means that the hydrogen bonding at this position, if it always exists, is not always "Watson-Crick." Crick (1966a) has proposed a less specific scheme of "wobble" hydrogen bonding, and this raises the more general question: Does *complete information* reside in the Watson-Crick hydrogen bonds between the codon on mRNA and the anticodon on tRNA, or does the selection of a particular tRNA depend also on the entire *biological context* in which the codon finds itself? Various conditions such as ion concentration, temperature, and the presence or absence of the 5' external phosphate group are known to affect "codon recognition" *in vitro*, and molecules such as streptomycin which alter the ribosome can change the meaning of codons both *in vitro* and *in vivo*.

If we depart from the concept that a particular codon always carries complete information for the specification of a particular amino acid, we are forced to adopt some type of linear programming hypothesis. We shall explore this speculative topic in Chapter 6 on the genetic code.

HOW THE TRANSMISSION IS CONTROLLED

Not only do we know where the information is stored and many of the mechanical details of its transmission; we also know that this transmission is governed by very delicately balanced control mechanisms, as we would expect in a complex information processing system. Whereas the evidence for the transmission mechanics is primarily chemical, depending ultimately on the chemical isolation and character-ization of each component, the evidence for the control mechanisms has been primarily genetic. However, with the isolation of repressors and other control molecules, this evidence will also ultimately become biochemical.

From biochemical studies, particularly of bacterial cells, we know that many copies of some proteins are present in the cell whereas others exist in trace amounts, perhaps some as only one molecule per cell. Further, the amounts of proteins present vary under certain environmental conditions.

From a detailed study of a body of genetic evidence, Jacob and Monod (1961) have constructed the following model to explain how protein synthesis is controlled. We will define their terminology as we go along. Substrates whose presence in the growth medium specifically increases the amount of an enzyme are called *inducers* and the enzymes, *inducible enzymes*. Biosynthetic enzymes whose amount is reduced by the presence of the end product molecules are termed *inducible enzymes*. The end product metabolites which specifically decrease the amount of a particular enzyme are called *corepressors*. Both inducers and corepressors combine with a regulatory group of protein molecules called *repressors*. The genes in DNA which code for the synthesis of repressors are called *regulatory genes*.

Repressors are highly specific molecular keys which lock and unlock the synthesis of specific proteins or groups of proteins. It is believed that they do this by combining specifically with a site on DNA called an *operator*, which controls the synthesis of a single mRNA molecule. The sequence of DNA bases coding for this single mRNA molecule is called an *operon*. Necessarily, the operator region immediately precedes the operon it controls. In fact, there is evidence for a region immediately preceding the operator called the *promotor* region (Jacob *et al.*, 1964). The promotor region exerts a still higher level of control in a system which is obviously a structured hierarchy of control levels. The hierarchy concept is so fundamental to the description of the living system that we shall refer to it often.

Not all proteins in the cell are under this control system. Proteins which are synthesized in fixed amount independent of need are called *constitutive* proteins. It is not known whether repressors themselves are constitutive proteins. If not, then their synthesis must be regulated by a higher-level control system which is entirely unknown at present.

A lower-level control system which is independent of repressor activation or inactivation is simple *feedback inhibition* wherein the end product of a synthetic pathway deactivates one of the enzymes in the pathway by combining directly with it. Again, we see the pattern of an ascending hierarchy of control mechanisms. Britten and Davidson (1969) have extended the model of Jacob and Monod in a model involving batteries of repeated functional units.

Jacob (1966) has constructed an interesting analogy between his control model and an electrical system. The analogy is somewhat intricate. One must "sit with it" to appreciate it, but the effort is well worthwhile because it is an excellent example of the principle that basic theories such as the model of Jacob and Monod are so general that they can be expressed in many formats.

In higher organisms, chromosomes come in homologous pairs, one chromosome of each pair coming from each parent. This is the diploid set, and the germ cell contains half this amount (the haploid set). In diploid or partially diploid microorganisms two classes of mutants can be isolated which map according to standard genetic techniques in the two different regions of the chromosomes corresponding to the regulator gene and the operator region. Regulator mutations are recessive to the wild type or natural state; however, operator mutations are dominant over wild type and affect only those genes *cis* to them, i.e., those on the same chromosome. Jacob (1966) draws an analogy between this situation and the electronic system in which two transmitters, t_1 and t_2, are sending out signals which keep two switches, s_1 and s_2, closed (s_1 and s_2 normally tend to remain open). If one of the transmitters is damaged, both switches are still regulated by the signal of the remaining undamaged transmitter and remain closed. However, if one of the switches becomes damaged so that it can no longer respond to the signal, the switch opens and is no longer subject to control. The molecular system corresponding to this analogy is shown in Fig. 4.

It is believed that in inducible systems the repressor is normally combined with the operator, thus preventing the transcription of the operon by mRNA. Hence the regulator gene is analogous to the

FIGURE 4

OPERON MODEL.

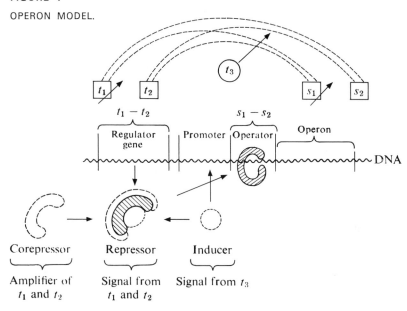

transmitter, the repressor is the signal, the operator is the switch. If in a diploid organism a mutation occurs in one of the chromosomes in the regulator region, the remaining homologous chromosome can still specify the synthesis of enough repressor to combine with both operators and the system remains subject to the same control mechanisms, i.e., the mutation is recessive to wild type. This is equivalent to damaging one of the transmitters. However, if in a diploid organism a mutation occurs in one of the operators so that it can no longer combine with the repressor or can no longer initiate the synthesis of mRNA, then all the genes *cis* to the mutation are affected. Systems which were formerly inducible become constitutive, i.e., the mutation is dominant over wild type. This corresponds to the damaging of one switch.

The normal functioning of this system depends on the interaction of the repressor with the inducer and corepressor. In inducible systems it is believed that the inducer combines with the repressor, changing its three-dimensional stereochemical configuration in such a way that its affinity for the operator is reduced. This is equivalent to blocking the signal from a normal transmitter by means of another signal from a third transmitter, t_3. The switch is then opened and mRNA synthesis begins. In repressible systems it is thought that the corepressor combines with the repressor, changing its configuration such that it has a greater affinity for the operator. This is equivalent to amplifying the output of a transmitter. This entire analogy is a magnificent demonstration of the fact that molecular biological models can be shown to be logically equivalent to other kinds of models, in this case, an electrical system.

Thus from chemical and genetic evidence biologists have constructed a fascinating description of the living system as an information processing machine with delicately balanced, intricate mechanisms of self-control. If we wish to understand this information processing machine deeply, we must understand the mathematical laws and operational principles which govern the storage and processing of information itself. Up to the present time, biologists have concentrated their efforts on describing the *mechanics* of control and the *mechanics* of information processing. It is time to call attention to more theoretical considerations which are of at least equal importance.

THE REDUCTIONIST—ANTIREDUCTIONIST CONTROVERSY

Because of the success of the Watson-Crick model of DNA structure in explaining so many of the features of the living system, because of this impressive body of knowledge about the mechanics of information processing and control, and because of the recent elucidation of the genetic code, many biologists, perhaps most, feel that all of life can be reduced to the "laws of physics and chemistry." Perhaps the most emphatic spokesman of this reductionist view is Francis Crick. In his book, *Of Molecules and Men*, Crick (1967) states: "So far everything we have found can be explained without effort in terms of the standard bonds of chemistry—the homopolar bond, the van der Waals attraction between non-bonded atoms, the all-important hydrogen bonds, and so on."

What such a view essentially says is that quantum mechanics is the underlying theoretical basis of life because all of chemistry can in principle be deduced from quantum theory. If this is the case, then the mathematical theory underlying biology now exists and all that is left to do is to work out the application as far as we can before encountering the barriers which arise only as a consequence of the complexity of the system.

We are observing today a most fascinating chapter in the history of science. At a time when biologists are particularly afflicted with this "all that is left to do is fill out the handbook" malady, a number of reputable physicists have stepped forward in direct opposition to it. The following quotation from Pattee (1967) summarizes the situation:

> Although the chemical bond was first recognized and discussed at great length in classical terms, most physicists regarded the nature of the chemical bond as a profound mystery until Heitler and London qualitatively derived the exchange interaction and showed that this quantum mechanical behavior accounted for the observed properties of valency and stability. On the other hand, it is not uncommon to find molecular biologists using a classical description of DNA replication and coding to justify the statement that living cells obey the laws of physics without ever once putting down a law of physics or showing quantitatively how these laws are obeyed by these processes.

As we have noted, the law that must be put down and demonstrated as relevant to the living systems is quantum mechanics. Von Neumann (1955) has proved that this area of knowledge forms a complete mathematical system which is unique under its postulates. Two able physicists, Elsasser (1958, 1966) and Wigner (1967), have attempted to find out by the responsible techniques which Pattee outlines if the living system does in fact obey this fundamental "law of chemistry and physics." However, both have met with serious difficulties. Neither Elsasser nor Wigner believes that quantum mechanical laws are adequate to explain the living system. Elsasser (1958) has postulated the existence of higher "biotonic" laws which govern living systems.

Wigner has calculated that the probability of existence of reproducing states is zero. He assumes a finite space instead of a Hilbert (or infinite) space and a random symmetric matrix for the interaction of the organism and the nutrient. Wigner is careful to point out that his model may not be applicable to the living system. However, he freely admits his "firm conviction in the existence of biotonic laws."

The reductionist, then, is one who believes that the theory is now complete and all of life can be reduced to the known "laws of physics and chemistry." At the other pole of the controversy is one who believes that the discovery of new laws will be necessary to explain the living system. Precisely what these new laws are no one seems to have stated clearly as yet. Perhaps this is why Crick (1967) labels such a person a "neo-vitalist—one who holds vitalistic ideas but does not want to be called a vitalist." But he misses the point. What these physicists are saying is that the presently existing laws of physics and chemistry may well turn out to be inadequate in the description of the living system for the same reasons that the laws of Newtonian mechanics were inadequate in dealing with the interior of the atom. This statement does not in any way imply a belief in the supernatural. The laws of Newton were adequate in dealing with the interactions of macroscopic bodies but, when Bohr attempted to use them in constructing his model of the atom, the result was quite disappointing, particularly when one attempted the quantitative explanation of spectra. It now begins to appear that quantum mechanics stands in a similar relationship to biology today.

We shall label the individual who disagrees with the reductionists simply an antireductionist. The most unique and forceful spokesman in my opinion for the antireductionist viewpoint is Michael Polanyi (1967). Polanyi draws directly on the laws of information theory in his attempt to explain why quantum mechanics is inadequate to explain the living system. He states that "all objects conveying information are irreducible to the terms of physics and chemistry." He believes that, because the living system is a machine, a concept often used by the reductionist, this is precisely why it is irreducible to the "laws of physics and chemistry." The reason is that we cannot understand an information processing machine, or any machine for that matter, merely from a description of its hardware. Only the systems expert really goes into a detailed study of hardware anyway. The vast majority of people who use computers study only the operational principles governing the behavior of the machine and, above all, the languages which one can use in conversing with it.

Polanyi points out that there are higher operational principles governing the design and function of a machine which cannot be deduced from a description of its hardware no matter how accurate and detailed this description might be. Also the operation of a machine is governed by boundary conditions which are not in any way determined by the structural laws of physics and chemistry. We shall discuss this in more detail in Chapter 5. These operational principles and boundary conditions constitute a more fundamental definition of the machine than its mere hardware and circuitry. There is a hierarchy of control whereby these higher operational principles take precedence over the lower "laws of physics and chemistry" which oversee the operation of the hardware in much the same manner as does a technician whose responsibility is restricted to keeping the machinery operating. Further, Polanyi believes that the higher operational principles fix the boundary conditions. This is a unique statement of the antireductionist position. It makes an attempt to outline what these "new laws" beyond the presently known "laws of physics and chemistry" might be.

Both parties to the controversy seem to agree that the living system may be regarded as a machine that stores and processes information. The reductionist can be likened to a systems expert whose objective is a detailed description of the computer's hardware. Such knowledge is certainly indispensable, and it is only natural that this is the first aspect of the living system to occupy our attention. However, the unfortunate belief prevails among biologists today that a nuts-and-bolts approach is sufficient to explain the living system. This unhealthy state of affairs is due in no small measure to a paper published in 1957 by Crick and co-workers. It contains a model which came to be known as "The Magic Twenty"; this model has done more harm to biology than any other single contribution in its history. The harm was, of course, unintentional; but, in my opinion, the circumstances were unfortunate. The model showed how one could code for an alphabet of 20 letters from another alphabet of 4 letters after making certain assumptions. Since DNA contains 4 bases and proteins contain 20 amino acids, the model was rapidly and widely accepted along with its assumptions. When the experimental techniques for actually delineating the genetic code showed this model to be entirely false, biologists became disenchanted with "theory" in general. Physicists, on the other hand, are not particularly impressed with the biologist's proclamation that theory is dead.

MATHEMATICS AND THE LAWS OF PHYSICS

I cannot state my position in this controversy without first defining some terms. The heart of the matter lies in the usage of the inexact phase, "the laws of physics and chemistry." If one includes information theory, game theory, and related areas of mathematical knowledge within the scope of this phrase, then I have reductionist leanings. However, I hardly see how we can make the scope of the phrase so broad at present. The laws of physics, like living organisms, are constantly evolving, and the laws of today that are "true" will take their place as special restricted cases under the more general laws of tomorrow. It is again an ascending, open-ended hierarchy.

Theoretical physics consists of the application of mathematical knowledge to the understanding of the physical universe. The undergraduate student begins by applying calculus in Newtonian mechanics. In quantum mechanics he applies his knowledge of the solution of partial differential equations. That particular portion of the body of mathematical knowledge which is applied to the description and prediction of the physical system becomes a part of the laws of physics. *In general, one does not consider unapplied mathematical knowledge a part of these laws.* For example, the tensor calculus first developed by Ricci and Levi-Civita was not a part of the "laws of physics" until Einstein applied it.

If the theoretical foundations of biology are incomplete, then there exists a body of mathematical knowledge, developed or undeveloped, which awaits application to the living system. I believe that information theory, game theory, and related areas constitute such a body of knowledge. This area of knowledge is still expanding and will perhaps all be unified some day. Parts of it such as game theory are admittedly incomplete. Even the genius of von Neumann (1947) made only a beginning in the area of non-zero-sum games, which are clearly of the greatest potential practical application. I see this general area of mathematical knowledge as part of "the laws of physics" some day; but we cannot reasonably consider it as such now because it is largely unapplied and even incomplete in its mathematical development.

Hence if we consider the scope of the phrase, "the laws of physics and chemistry," to be confined to mathematical knowledge which has already been applied in detail to the physical universe, and particularly if we restrict our thinking to such narrow phrases as "the standard bonds of chemistry"; then I am an antireductionist. I hope that this work will be a first step in the detailed application of information theory to the living system. Conceptually only one application has been made up to the present time; it is as follows. We know that the maximum information content that DNA can have, since it is based on an alphabet of 4 letters, is 2 bits per symbol. Knowing the amount of DNA in a particular organism, we can estimate the maximum amount of information its DNA can contain. We then ask the question: "Is this enough information to specify the structure of the mature organism?" In answering this question, one can only guess at how much information is represented by the three-dimensional structure of the organism. Various estimates have been made (Dancoff and Quastler, 1953). Bremermann (1967) presents a model of the brain as a wiring board which shows that, if each connection must be specified, there is not enough information in DNA even to construct the vertebrate brain, let alone all the rest of the animal.

One explanation is that the organism is constructed of simple repetitive units which are self-assembling to some extent, so that we do not need nearly so much information to specify structure as the estimates based on combinatorial formulas would lead us to believe. An alternative and related explanation is that there is a hierarchy of instructions or programs for the construction of living systems and only the highest-level program, the "master program," need be stored since it can generate all subprograms. A simple analogy is a geometry textbook. The entire book can, in principle, be generated from a few simple axioms and postulates.

This is an interesting problem, but it is characterized by inexactness and approximation. One of the most universal characteristics of the process whereby mathematical knowledge becomes a "law of physics" is the description of a system in such exact numerical terms that quantitative prediction of experimental fact inevitably follows. One can hardly describe the work mentioned above in such terms. Apter and Wolpert (1963) made essentially the same criticism of all such work. Specifically, they point out that what is needed is a theory which will tell us exactly how much information a given DNA sequence actually contains. Such a theory is one of the contributions of this work.

HOW MUCH INFORMATION IS STORED

Perhaps the most primitive quantitative questions man can ask are "how many?" in the case of a discrete variable and "how much?" in the case of a continuous variable. Since information is a continuous variable, we begin with the primitive question: "How much information is stored in the base sequence of a given DNA molecule?"

Here we meet the first question before which the structural reductionist is completely mute, provided, of course, that he understands the question. The question is "how much?" and can only be answered with a number. If he does not understand the question, he will attempt to answer by describing the transmission mechanics, the control mechanics, the genetic structure of the organism, etc., but no amount of structural knowledge alone, no matter how detailed, can answer this question, not even a complete knowledge of the total base sequence of every DNA molecule in existence. This question has no meaningful answer in terms of "the standard bonds of chemistry," be they homopolar, van der Waals' or Watson-Crick. This is the point Polanyi (1967) is making when he argues that the base sequence of a particular DNA molecule is not determined by chemical forces; yet it is precisely in the unique sequential arrangement of the bases that the information resides. Before we can even begin to consider the evolutionary question of the forces which selected the presently existing sequences, we must first answer the primitive questions. They were, after all, the questions which gave rise to information theory.

Unlike the tensor calculus, which was first developed and then applied, information theory arose in response to the need for a mathematical framework within which to fit a physical situation. It arose from the needs of electrical engineers working on communications problems. The engineer needs to know *how much* information is being transmitted over a particular communications system, such as a telegraph system, in order to determine its efficiency. He needs to know if a particular code, such as the Morse code, transmits information more efficiently than another. In 1949 Claude Shannon of Bell Telephone Laboratories published his classic paper, entitled "The Mathematical Theory of Communication" (in Shannon and Weaver, 1949), which is often said to have "founded" the field of information theory. Later, mathematicians were able to show that information theory is, in fact, a body of pure mathematical knowledge, i.e., it can be derived from first principles and rests largely on the foundation of probability theory. Shannon's paper, which was a singular contribution to the fields of physics and engineering, was in many respects incomplete, inexact, and even incorrect mathematically. Khinchin (1957), the Russian mathematician who gave one of the first complete treatments of information theory, states that "besides having to free the theory of practical details, in many instances I have amplified and changed both the statement of definitions and the statement and proofs of theorems."

This is not to devaluate Shannon's contribution. But, as so often happens, broad theoretical principles arise within restricted physical settings. Hence their terminology and even methodology are subjected to artificial constraint. The field of information theory even in its broadest mathematical statement still carries some of the trappings of the communications engineer. I shall show that the biologist and the engineer are interested in different aspects of the information concept. This is a fundamental reason why information theory has not been broadly applied to biology.

Even the viewpoint and statement of the primitive question are different. The communications engineer essentially asks "How much information can be transmitted?" but, as a biologist, I must first ask "How much information can be stored?" I shall elaborate on this point later but it must be stated now. Because of the subtle difference in the primitive question, I take a somewhat different route in the derivation of the mathematical expressions. The functional forms turn out to be analogous to those of classical information theory but the interpretation is different. Also there is a subtle but all-important difference in the basic equations. This result, which arises from the difference in primitive starting point of the theory, is of the utmost importance in describing living systems. In fact, we may say that the need to understand the living system has given rise to this extension of the theory.

The unifying thread of our story is the entropy concept. As *Homo sapiens*, we have always believed that we are higher organisms. After all, we are more complex, more differentiated, more highly ordered than lower organisms. As thermodynamicists, we recognize these words and realize that the concept of entropy must somehow enter into our explanation. We have always had the vague notion that, as higher organisms have evolved, their entropy has in some way declined because of this higher degree of organization. For example, Schrödinger made his famous comment that the living organism "feeds on negative entropy."

We reason that this decreasing entropy of evolving life, if it exists, does not in any way violate the second law of thermodynamics, which states that the entropy of an isolated system never decreases. The living system is not isolated and the reduction in entropy could have been compensated for by a correspondingly greater increase in the entropy of the surroundings. It does not violate the letter of the second law, and yet something about it seems to make us uneasy. Why should the evolution of the living system constantly drive in the direction of increasing organization while all about us we observe the operation of the entropy maximum principle, which is a disorganizing principle? I know of no other system except the living system that does this.

First of all, can we establish that the entropy has in fact declined in higher organisms? No one has ever proved this quantitatively. In fact, one can argue that it is impossible to establish this thesis by classical means because of the uncertainty principle in its broadest sense. In particular, if we were to make the precise and extensive measurements necessary to determine accurately the entropy difference between a higher and a lower organism, these measurements would disturb the living systems so much that they would kill them. Thus it is impossible by classical means even to establish this proposition in which almost all of us seem to believe.

When concepts break down like this, they are of little use to us. I think our classical notions of entropy as they come to us from the presently established "laws of physics and chemistry" are totally inadequate in dealing with the living system. This does not mean that there is anything mysterious, supernatural, or vitalistic about the living system. It simply means that our classical notions are *inadequate*.

I shall extend the entropy concept primarily through the apparatus of information theory, but I shall extend that also. Shannon (1949) gave the most general definition of entropy to date, and I shall extend his concept. Specifically, I shall show that the entropy function which Shannon called the redundancy is composed of two parts which I call

D_1 and D_2. We must characterize the redundancy of a sequence of symbols by two independent numbers, one describing the amount and the other the kind of redundancy of the sequence. I can state this in terms of entropy. I shall show that phrases like increasing entropy or decreasing entropy are not completely definitive. We must ask, "In what way has the entropy increased or decreased or what *kind* of entropy is it?" We do not encounter such questions in either classical thermodynamics or information theory. I shall develop a theory which can answer these questions.

In classical thermodynamics we dealt with the ordering of three-dimensional aggregates of matter, but in information theory we begin to grapple with the concept of ordering of one-dimensional sequences of symbols. This is very significant because we now know that the DNA molecule is a linear sequence of symbols which stores the primary hereditary information from which the entire living organism is derived, just as a set of axioms and postulates stores the primary information from which a mathematical system is deduced. Therefore, if we wish to investigate the organization of living systems, we must investigate the ordering of the sequences of symbols which specify them.

BACKGROUND BOOKS

Abramson, N. (1963). *Information Theory and Coding*, McGraw-Hill, New York.

Ash, R. B. (1965). *Information Theory*, McGraw-Hill, New York.

Brillouin, L. (1956). *Science and Information Theory*, Academic Press, New York.

Cherry, C. (1957). *On Human Communication*, Technology Press of Massachusetts Institute of Technology, Cambridge, Mass.

Crick, F. H. C. (1967). *Of Molecules and Men*, Univ. of Washington Press, Seattle, Wash.

Elsasser, W. M. (1958). *The Physical Foundations of Biology*, Pergamon Press, New York.

Elsasser, W. M. (1966). *Atom and Organism*, Princeton Univ. Press, Princeton, N.J.

Hartman, P. E., and Suskind, S. R. (1965). *Gene Action*, Prentice-Hall, Englewood Cliffs, N.J.

Jukes, T. H. (1966). *Molecules and Evolution*, Columbia Univ. Press, New York.

Khinchin, A. I. (1957). *Mathematical Foundations of Information Theory*, Dover, New York.

Koestler, A. (1967). *The Ghost in the Machine*, Macmillan, New York.

Polanyi, M. (1966). *The Tacit Dimension*, Doubleday, New York.

Quastler, H. (1964). *The Emergence of Biological Organization*, Yale Univ. Press, New Haven, Conn.

Reza, F. M. (1961). *An Introduction to Information Theory*, Wiley-Interscience, New York.

Shannon, C. E., and Weaver, W. (1949). *The Mathematical Theory of Communication*, Univ. of Illinois Press, Urbana, Ill. (paperback, 1963).

Singh, J. (1966). *Great Ideas in Information Theory, Language and Cybernetics*, Dover, New York.

Stent, G. S. (1963). *Molecular Biology of Bacterial Viruses*, W. H. Freeman and Co., San Francisco, Calif.

Watson, J. D. (1965). *Molecular Biology of the Gene*, W. A. Benjamin, Inc., New York.

Wigner, E. P. (1967). *Symmetries and Reflections*, Indiana Univ. Press, Bloomington, Ind.

Yčas, M. (1969). *The Biological Code*, North-Holland Publ. Co., Amsterdam.

2 *The obvious is that which is never seen until someone expresses it simply.*
—*Kahlil Gibran, Sand and Foam*

ENTROPY IS THE MEASURE

INFORMATION IS A CAPACITY

What is information? To be honest, information is an ultimately indefinable or intuitive first principle, like energy, whose precise definition always somehow seems to slip through our fingers like a shadow. Webster essentially defines information as knowledge and knowledge as information. This irreducibility, however, does not mean that we cannot define information *operationally* as we do energy and understand a great deal about its nature and how it expresses itself in the world about us. We often define energy operationally as the capacity to do work, not the work itself. Similarly we may define information operationally as the capacity to store and transmit meaning or knowledge, not the meaning or knowledge itself. The idea of capacity is the key that opens the door to a quantitative formulation of the concept.

Suppose I had two books with identical information contents, one in English and the other in Chinese. The English book would convey meaning only to the English-speaking person and the Chinese book to the Chinese-speaking person, but the *capacity* to store and transmit this meaning would be the same for both. Thus, in the technical sense, the word information, like energy, is a capacity, and one must be constantly on his guard to distinguish between the scientific meaning of the word and its common usage as a synonym for knowledge, meaning, value, etc. One cannot avoid the latter even in scientific discussion. However, when we calculate a numerical value for the information, it is a capacity we calculate—always. How do we calculate this capacity? Shannon was the first to suggest that *entropy is the measure* of this capacity and to develop the specific functional form of this measure.

SOME DEFINITIONS

Entropy measures the degree of randomness of a system. Therefore in order to define entropy we must define randomness, and this involves the concept of probability. We begin with some fundamental definitions.

A *random phenomenon* is a group of related but not always identical occurrences wherein the individual outcome cannot be predicted or predetermined in any way but occurs according to chance, whereas a large number of these related outcomes fit a pattern that can be predicted quite accurately. For example, we are all familiar with the bell-shaped Gaussian curve. If we give a standard intelligence test to a large number of college freshmen, we cannot predict exactly what a particular student will score, but we can predict that all the scores will fall into a bell-shaped pattern. A *random event* is a single outcome or type of outcome of a random phenomenon such that its relative frequency of occurrence approaches a stable limiting value in a large number of possible occurrences. The limiting value itself is called the *probability* of the random event. Thus the probability concept involves more than merely the idea of a relative frequency of occurrence. It involves the idea of an infinite series and its limiting value. For example, if I toss a fair coin two times, I do not necessarily obtain one head and one tail. However, if I toss the coin a million times, the relative frequency of occurrence of either heads or tails will approach exactly one-half, coming closer and closer to this value as the number of trials increases indefinitely.

A *set* is a collection of individual objects or ideas under study and is denoted by brackets. For example,

$$\{1, 2, 3\}, \quad \{1, 3, 2\}, \quad \text{and} \quad \{1, 1, 2, 3\}$$

are all the same set since they are all collections of the same individual entities, 1, 2, and 3. Identity of individuals is thus the dominant idea underlying sets; hence it is superfluous to name the same individual twice when defining a set. The order in which the individuals are written is likewise immaterial. The individuals themselves are called *elements* or members of the set. A *space* is a set that is in some way complete so that only those elements which "belong" and all of them that "belong" are included. For example, a space might consist of all the real numbers, but we would not include a horse in this space. Neither would we omit the number three. A space is a complete set. A *sample description space* is a space whose elements are the descriptions of all possible outcomes of a random phenomenon, and each element is called an elementary random event. With every element of the space is associated a number, the probability of the elementary event, which lies somewhere between 0 and 1; and the sum of all the probabilities is 1. Such a space is also called a *finite probability space* when the number of elements is finite. For example, the sample description space of the random phenomenon of tossing a coin is

$$S_1 = \{H, T\}$$

where H is heads and T is tails. For tossing two coins consecutively,

$$S_2 = \{HT, TH, HH, TT\}$$

Two random events, *a* and *b*, are said to be *independent* if the probability of their joint occurrence *ab* is the product of the probabilities of their separate occurrences, i.e.,

$$p(ab) = p(a)p(b) \tag{1}$$

For example, the tossing of one coin does not in any way influence the outcome of tossing a second. Therefore the events are independent and the probability of tossing two heads is

$$p(HH) = p(H)p(H) = \tfrac{1}{2} \times \tfrac{1}{2} = \tfrac{1}{4}$$

This is intuitively obvious simply from examining the sample description space S_2. The elementary events are all equally probable and the joint occurrence of two heads represents one description out of four.

Two random events are said to be *dependent* if the previous occurrence of one alters the probability of occurrence of the second. For example, if I have an urn containing three white and three black balls, the probability of drawing either a white or a black ball from the urn is one-half. However, if I draw one white ball from the urn without replacing it, then the probability of drawing a white ball on the second draw is two out of five and the probability of drawing a black ball is three out of five. These altered probabilities are called *conditional probabilities*, and we must always specify what has altered them, i.e., the previously occurring event must be given. If we have any two dependent random events, *a* and *b*, the probability of their joint occurrence, *ab*, is the probability of *a* multiplied by the conditional probability of *b*, given that *a* has occurred. This is written

$$p(ab) = p(a)p(b \mid a) \tag{2}$$

For example, the probability of drawing two white balls from the urn without replacing the first one is

$$p(ww) = (\tfrac{1}{2})(\tfrac{2}{5}) = \tfrac{1}{5}$$

ELEMENTS OF THE ENTROPY CONCEPT

Scrabble is a game that is played with small pieces of wood or cardboard on one side of which is imprinted a single letter from the English alphabet. If I take the pieces from a Scrabble game and mix them up in a completely disordered pile, we characterize this state of the system with words such as random, disordered, disorganized, mixed, homogeneous. If entropy measures the randomness of a system, this is clearly a state of high entropy. If I take each piece and turn it face up on the table, the entropy is lowered and, if I separate the letters into two groups, one containing only vowels and the other only consonants, the entropy has been lowered further. With the state of lower entropy we associate words such as nonrandom, ordered, organized, separated, inhomogeneous. Table 1 lists these concepts. They are intuitive, qualitative concepts. How can we make them quantitative?

Let us again scramble the pieces into a jumbled heap. If I reach without looking and draw a piece from the pile, then note whether it was face side up or blank side up, my chances of drawing a blank or a letter are about equally probable. In fact, we feel that, if the pile were very large and the pieces were "perfectly" mixed, they would be exactly equiprobable. We thus arrive intuitively at the notion that the maximum entropy state is characterized by equiprobable elementary random events. This is a fundamental quantitative concept. Note that we arrived at this notion by imposing a random sampling operation on the system.

TABLE 1

ELEMENTS OF THE ENTROPY CONCEPT

Higher Entropy	*Lower Entropy*
Random	Nonrandom
Disorganized	Organized
Disordered	Ordered
Mixed	Separated
Equiprobable events	D_1 (divergence from equiprobability)
Independent events	D_2 (divergence from independence)
Configurational variety	Restricted arrangements
Freedom of choice	Constraint
Uncertainty	Reliability
Higher error probability	Fidelity
Potential information	Stored information

The intuitive idea that the most random state is characterized by equiprobable events is illustrated in many games. For example, a "fair" die is one that is perfectly balanced so that all sides have an equally probable chance of appearing on a given roll, a "fair" roulette wheel is one where all the numbers have an equally probable chance of being selected on a given spin of the wheel, etc.

The pieces of wood in the Scrabble game have no affinity for each other. Where one occurs in the pile has no effect on where another will occur, i.e., they are independent events. However, if there were some attractive force between the e and i, the t and h, etc., this would introduce an ordering force into the system and the entropy would be lowered. Hence with the maximum entropy state we also associate the idea of independent elementary events. We can now state a fundamental, quantitative principle. The *maximum entropy state is characterized by equiprobable, independent elementary events.*

GAS IN A BOX

There are other important elements of the entropy concept that can best be illustrated with the classic thermodynamic example of a gas in a box. Suppose we have two distinguishable gases, A and B, separated by a removable partition in a thermally insulated container. We note that this is a state of some separation and ordering since each gas is confined to only half of the box. What will happen if we remove the partition? We all know that the gas molecules will interdiffuse and intermix until a homogeneous state is reached. The entropy has increased. We also know that they will never spontaneously separate again with gas A going back to one side and gas B to the other. We are here observing the second law of thermodynamics, namely, that the entropy of an isolated system never decreases; it either increases or remains constant. We sometimes say that the entropy of an isolated system seeks a maximum.

Let us divide our box into imaginary compartments of molecular dimensions and note which individual molecules are in a particular compartment at a given time. Physicists call such a specified arrangement of the system a "microstate." Since the molecules of a gas are constantly in rapid, random motion, colliding and rebounding like billions of tiny billiard balls, different molecules will occupy different compartments at different times, and many arrangements or microstates of the system are possible. If we count the number of molecules in each compartment, this set of numbers is said to define a "macrostate" of the system. The exact number of microstates possible for a given macrostate bears the impressive name of the "thermodynamic probability" of that macrostate and is denoted by the letter W.

Before removal of the partition, the molecules of gas A are confined to one side of the box and the molecules of gas B to the other. A little reflection will convince us that after removal of the partition a greater number of microstates is possible because every molecule is now free to roam over the entire box. There is greater uncertainty as to where a particular molecule will be than there was before removal of the partition. Thus with the state of higher entropy we associate the concepts of greater freedom, uncertainty, and more configurational variety; and the quantitative expression of these concepts is that the number of microstates has increased. When we consider the entropy of a linear sequence of symbols ordered according to the constraints of a language, the concept of configurational variety means greater word or message variety.

The elements of uncertainty and configurational variety will be especially important when we discuss the interrelationship of the entropy and information concepts. The concept of uncertainty implies a higher probability of error. For example, if I had to guess where a particular molecule of gas A would be at a given time, my probability of error would be higher in the higher entropy state. Before removal of the partition we at least knew that it was on a particular side of the box. The constraint and ordering of the lower entropy state carries with it a certain reliability and lower probability of error. This is a central concept in information theory.

At this point we should unify the words that have proliferated in Table 1. We have really derived only two basic concepts from our consideration of the gas in the box and the Scrabble game. They may be expressed in a variety of ways such as freedom versus constraint, variety versus reliability, uncertainty versus fidelity; but there are basically only two conceptual elements. A higher number of microstates is simply a quantitative expression of greater configurational variety. This in itself necessitates greater freedom of arrangement or movement of the particles, which in turn means greater freedom of choice, greater uncertainty, and higher probability of error in the prediction of the outcome of a random sampling operation. On the other side of the table, reliability or fidelity is the consequence of the restriction, constraint, and certainty associated with the lower entropy state. We shall find that reliability and variety are the two mutually antagonistic elements, the thesis-antithesis, which constitute the essential nature of the information concept.

QUANTITATIVE DEFINITION OF ENTROPY

It is clear from the preceding discussion that, as the number of microstates increases, the entropy increases. We might at first attempt to express this relationship as a direct proportion. Hence for a given macrostate we might write

$$S = KW \tag{3}$$

where S denotes the entropy, W is the thermodynamic probability, and K is an arbitrary constant. However, this definition is not yet complete or even correct. In addition to the fact that entropy is a monotonically increasing function of W, it is intuitively reasonable that entropy should also have the additive property, i.e., the entropy of system A plus the entropy of system B should be equal to the entropy of the composite system AB.

$$S_A + S_B = S_{AB} \tag{4}$$

For example, if the entropy of system A increases by two entropy units and the entropy of system B by three entropy units, we feel intuitively that the entropy of the universe has increased by five entropy units, other things being constant. There is no reason to feel that the increase in one system should in any way offset the increase in another. Thus we feel that the quantitative definition of entropy should combine properties (3) and (4); yet it is not immediately obvious how this can be done since the number of microstates of composite system, AB, is the product *not* the sum of the microstates of system A and system B.

Boltzmann solved this problem in the early eighteen hundreds. When numbers expressed as powers of the same base are multiplicative, their exponents are additive. Logarithms are such exponents. Therefore we can combine both properties (3) and (4) in the definition

$$S = K \log W \tag{5}$$

Thus the entropy of system A may be written

$$S_A = K \log W_A \tag{6}$$

and, for system B,

$$S_B = K \log W_B \tag{7}$$

Adding these entropies, we have the entropy of the composite system,

$$S_{AB} = K \log W_A + K \log W_B \tag{8}$$

We conclude that

$$S_{AB} = K \log W_A W_B \tag{9}$$

or

$$S_{AB} = K \log W_{AB} \tag{10}$$

It works out: Boltzmann's definition has both of the desired properties.

SHANNON'S FORMULA

Statistical thermodynamics, which we have been studying, is based on the assumption that all microstates are equiprobable. If this is the case, then the probability of each individual microstate, p_i, is simply one out of the total number of microstates, W, i.e.,

$$p_i = \frac{1}{W} \quad \text{or} \quad W = \frac{1}{p_i} \tag{11}$$

Substituting this expression for W in equation (5),

$$S = K \log \frac{1}{p_i} \tag{12}$$

and, since the log of 1 is 0,

$$S = -K \log p_i \tag{13}$$

Keep in mind that equation (13) holds only for the case when the microstates or arrangements of a system are equiprobable. It is good, however, that we can express the entropy in terms of a probability rather than W because a large number like W is often impossible to determine anyhow, whereas a probability is often measurable experimentally. Would it be possible to express entropy in terms of probability even if the arrangements of the system were not equiprobable? It is in fact possible with the use of a statistical concept called the expectation value, or simply the statistical average.

Suppose we have a numerical-valued random phenomenon (i.e., with each outcome of the random phenomenon there is associated a number). For example, when we toss a die, we record the outcome as the number of dots on the face that lands up. We may ask, "What will be the average of all these listed numerical values as the number of outcomes increases indefinitely?" This is simply the sum of all possible numerical values, each one multiplied by the probability of its own occurrence. For example, if we toss a fair die and list each outcome, as the number of trials becomes very large the average numerical value of all the listed outcomes is

$$\tfrac{1}{6}(1) + \tfrac{1}{6}(2) + \tfrac{1}{6}(3) + \tfrac{1}{6}(4) + \tfrac{1}{6}(5) + \tfrac{1}{6}(6) = 3\tfrac{1}{2}$$

This is an intuitively reasonable result. If all sides of the die are equally probable, then the average numerical value should lie exactly midway between the 1 and the 6. However, if the die is loaded so that the 6 falls up 2/3 of the time and the 1 only 1/3 of the time, we would expect the average numerical value of the outcomes to be somewhat higher than this. It is:

$$\tfrac{2}{3}(6) + \tfrac{1}{3}(1) = 4\tfrac{1}{3}$$

Thus the expected value, or expectation value, of a numerical-valued random phenomenon is the sum over all possible outcomes of the probability of each individual outcome multiplied by the numerical value of that individual outcome. We may place this entire sentence in symbols.

$$E_x = \sum_i p_i n_i \tag{14}$$

With every arrangement of a system there is associated the number $-K \log p_i$, and the probability of each arrangement is p_i. The expectation value of this numerical-valued random phenomenon, which we will denote H, is

$$H = -K \sum_i p_i \log p_i \tag{15}$$

This is Shannon's formula. It is the expectation value of the Boltzmann variable, $-K \log p_i$. It expresses the entropy of a system in terms of probabilities and may be used even when all the microstates or elementary arrangements of the system are not equiprobable. It is a remarkable accomplishment; it takes the concept of entropy out of the restricted thermodynamic setting in which it arose historically and lifts it to the higher domain of general probability theory.

Shannon's formula can be derived by much more sophisticated mathematical techniques than we have used here. In our discussion of the elements of the entropy concept we showed intuitively that the maximum entropy state is characterized by equiprobable elementary events, i.e., all the p_i are equal. It can be shown rigorously that the function (equation 15) takes on its maximum value if and only if all the p_i are equal (Khinchin, 1957). Not only this; if we specify certain other additive properties, this function is unique. It is the only function possible which has the desired properties.

We must now discuss the matter of units. The logarithm base we use is a matter of arbitrary choice, and the value of the proportionality constant is also arbitrary. When K is set equal to 1 and base 2 logarithms are used, the unit of entropy is called a bit. When $K = 1$ and base 10 logs are used, the unit is called a Hartley. One Hartley = 3.3219 bits. Bits are the most generally used units.

THE DIVERGENCE FROM EQUIPROBABILITY

The power of Shannon's formula lies in its generality. Although we have developed it by using the example of a gas in a box and the concept of microstates, the p_i may refer to the probabilities of *any* elementary events defined on *any* sample description space. For example, let us define the sample description space S_1 of the random phenomenon of choosing by some suitable chance device a single base anywhere along a given DNA chain:

$$S_1 = \{A, T, C, G\} \tag{16}$$

where A is the base adenine, T is thymine, C is cytosine, and G is guanine. The entropy of this space is

$$H_1 = -K \sum_i p_i \log p_i \tag{17}$$

But the p_i are known. They are simply the base composition of DNA. For example, the base composition of *Micrococcus lysodeikticus* is

$$p(C) = p(G) = .355 \quad \text{and} \quad p(A) = p(T) = .145$$

Hence, for *M. lysodeikticus*,

$$H_1 = -(.355 \log .355 + .355 \log .355 + .145 \log .145$$
$$+ .145 \log .145) = 1.87 \text{ bits}$$

As a second example let us consider the DNA of *E. coli* where all the bases are equiprobable, $p(A) = p(T) = p(C) = p(G) = \frac{1}{4}$.

$$H_1 = -\log \tfrac{1}{4} = \log 4$$
$$H_1 = 2 \text{ bits}$$

Since all the bases are equiprobable, this must be the maximum value H_1 can ever have. We may express this in a more general way by letting a be the number of letters in the alphabet ($a = 4$ for DNA). When all the letters are equiprobable,

$$p_i = \frac{1}{a}$$

and

$$H_1 = -\log p_i = -\log \frac{1}{a} = \log a$$

It can be proved rigorously (Khinchin, 1957) that $\log a$ is the maximum value that H_1 can ever take on, i.e.,

$$H_1^{\text{Max}} = \log a \tag{18}$$

The divergence from this equiprobable state, which we will call D_1, is the maximum value H_1 can have minus the value it actually does have. In symbols,

$$D_1 = H_1^{\text{Max}} - H_1 = \log a - H_1 \tag{19}$$

For example, *E. coli* has no divergence from equiprobability because $H_1 = H_1^{\text{Max}}$ but, for *M. lysodeikticus*,

$$D_1 = \log a - 1.87 = 2.00 - 1.87 = .13 \text{ bit}$$

THE DIVERGENCE FROM INDEPENDENCE

It is easy to see that D_1 and H_1 tell us only part of the story. They are both based on S_1, a space of single-letter events, which contains no information about how these letters are arranged in a linear sequence. H_1 is a function only of the base composition of DNA, and D_1 tells us only how much of the total divergence from the maximum entropy state is due to the divergence of the base composition from a uniform distribution. We said that the maximum entropy state is characterized by equiprobable and independent events. Therefore we must now ask the question, "Are the bases in the DNA chain independent events?" Does the occurrence of any one base along the chain alter the probability of occurrence of the base next to it? Suppose the base A occurs. What is the probability that it will be followed by another A, or T, or C, or G? Will the probabilities of these bases be changed from their standard base composition values? In other words, we are asking "What are the numerical values of the conditional probabilities?"

$p(\text{A} \mid \text{A})$

$p(\text{T} \mid \text{A})$

$p(\text{C} \mid \text{A})$

$p(\text{G} \mid \text{A})$?

If we could measure these conditional probabilities in some way, and if we were to find that they were the same as the base composition, i.e.,

$$p(A \mid A) = p(A)$$
$$p(T \mid A) = p(T)$$
$$p(C \mid A) = p(C)$$
$$p(G \mid A) = p(G)$$

the bases would be independent of each other. However, if we were to find that the conditional probabilities were not the same as the base composition, then there would have been some divergence from independence of the bases. These conditional probabilities have, in fact, been measured for the DNA or RNA of more than 60 organisms and tissues by the nearest neighbor experiment of Kornberg's group (Josse *et al.*, 1961). In not a single DNA or RNA tested so far are the bases completely independent events, not even for *E. coli*, where they are equiprobable.

The divergence from independence results from a linear ordering of the bases along the chain, which reduces the entropy. In order to get some idea of how this may come about, let us give two examples of extreme cases of high divergence from independence even when the bases are equiprobable. Consider the sequence

AAAAA ... TTTTT ... CCCCC ... GGGGG ...

in which the dots signify lengthy extensions of the preceding pattern. Here the conditional probability of A given A is very high, of T given A very low but finite, whereas the conditional probabilities C given A and G given A are 0. This is a state of extreme ordering. There are other ways. Consider for example the sequence

ATCGATCGATCGATCGATCGATCG ...

Here $p(T \mid A) = p(C \mid T) = p(G \mid C) = p(A \mid G) = 1$ and all others are 0. We shall prove in Chapter 7 that the divergence from independence of this sequence is maximal even though the bases are equiprobable. We need a quantitative expression which will tell us exactly how much any given DNA base sequence has deviated from the maximum entropy state due to this linear ordering, which is a divergence from independence of the bases and expresses itself quantitatively as a deviation of the conditional probabilities from the base composition values.

We cannot define a linear ordering on a space of single-letter events such as S_1. We need at least a space of letter pair events, S_2. Therefore we define a sample description space of doublet or letter pair sequences as

$$S_2 = \{AA, AT, AC, AG, TA, TT, TC, TG, CA, \\ CT, CC, CG, GA, GT, GC, GG\} \quad (20)$$

The entropy of S_2 is

$$H_2 = -[p(AA) \log p(AA) + p(AT) \log p(AT) + \cdots] \quad (21)$$

But what is the probability of the doublet event? It depends on whether or not the single-letter events are independent or dependent. Recalling our original definitions, equations (1) and (2), if the single-letter events are independent,

$$p(AA) = p(A)p(A)$$
$$p(AT) = p(A)p(T), \quad \text{etc.} \quad (22)$$

but, if they are not independent,

$$p(AA) = p(A)p(A \mid A)$$
$$p(AT) = p(A)p(T \mid A), \quad \text{etc.} \quad (23)$$

The nearest neighbor experiment shows that the bases in DNA are not independent and it measures the doublet probabilities, which are called "nearest neighbor frequencies" in the original literature. Inserting these values in equation (21) gives the value of H_2 when the bases are dependent, which we will call H_2 dependent or H_2^D. We can always calculate what the value of H_2 would be *if* the bases were completely independent by using the base composition and equation set (22) to calculate the doublet probabilities. We will call this entropy H_2 independent or H_2^{Ind}. The divergence from independence, which we will call D_2, is simply the difference between the two.

$$D_2 = H_2^{Ind} - H_2^D \quad (24)$$

The sum of D_1 and D_2 is the total divergence from the maximum entropy state.

FIGURE 5

THE ENTROPY SCALE.

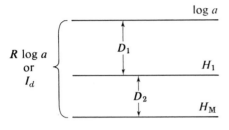

Figure 5 is an entropy scale. The scale runs from 0 to log a, the maximum value the entropy can have. D_1 is the distance log $a - H_1$, and we will later show that D_2 is the distance $H_1 - H_M$. H_M (or "H Markov") is the entropy value which takes into account both the distances, D_1 and D_2. We will describe it in detail later. For now all that is important is to realize that we have located a point on a relative entropy scale, and we may express this either as a distance from the zero point of the scale or from its maximum value.

A SAMPLE CALCULATION

Let us calculate D_1 and D_2 for *Micrococcus phlei* DNA. Its base composition is: $p(A) = .164$, $p(T) = .162$, $p(C) = .337$, $p(G) = .337$. From equation (17),

$$H_1 = -(.164 \log .164 + .162 \log .162$$
$$+ .337 \log .337 + .337 \log .337)$$
$$H_1 = 1.910 \text{ bits}$$

From equation (19),

$$D_1 = \log a - H_1 = 2.000 - 1.910$$
$$D_1 = .090 \text{ bit}$$

From equation set (22) and the base composition we can calculate the doublet probabilities as if the bases were independent events. They are shown in the accompanying tabulation. From it and from equation (21) we can calculate H_2^{Ind}.

$$H_2^{\text{Ind}} = -(.0262 \log .0262 + .0267 \log .0267 + (\cdots)$$
$$H_2^{\text{Ind}} = 3.819 \text{ bits}$$

	A	T	C	G
A	.0262	.0267	.0543	.0548
T	.0267	.0272	.0553	.0558
C	.0543	.0553	.1122	.1132
G	.0548	.0558	.1132	.1142

To calculate H_2^{D} we must know the dependent doublet probabilities. They were calculated and reported in the original literature (Josse *et al.*, 1961) as "nearest neighbor frequencies." For *M. phlei* the nearest neighbor frequencies are given in the accompanying tabulation. From equation (21),

$$H_2^{D} = -(.024 \log .024 + .031 \log .031 + \cdots)$$
$$H_2^{D} = 3.792 \text{ bits}$$

$\begin{array}{c} 5' \\ \nearrow \\ 3' \end{array}$	A	T	C	G
A	.024	.031	.064	.045
T	.012	.026	.061	.063
C	.063	.045	.090	.139
G	.065	.060	.122	.090

From equation (24),

$$D_2 = H_2^{\text{Ind}} - H_2^{D} = 3.819 - 3.792$$
$$D_2 = .027 \text{ bit}$$

and

$$D_1 + D_2 = .117 \text{ bit}$$

THE NEAREST NEIGHBOR EXPERIMENT

Arthur Kornberg was awarded the Nobel Prize for his discovery of the enzyme DNA polymerase, which synthesizes DNA, in the test

TABLE 2[a]

THE NEAREST NEIGHBOR EXPERIMENT

Experiment I with dATP³²

Monomer Isolated	Doublet Sequence	Cpm	Fraction
T	TpA	873	.075
A	ApA	1,710	.146
C	CpA	4,430	.378
G	GpA	4,690	.401
		11,703	1.000

Experiment II with dTTP³²

Monomer Isolated	Doublet Sequence	Cpm	Fraction
T	TpT	1,665	.157
A	ApT	2,065	.194
C	CpT	2,980	.279
G	GpT	3,945	.370
		10,655	1.000

Experiment III with dGTP³²

Monomer Isolated	Doublet Sequence	Cpm	Fraction
T	TpG	3,490	.187
A	ApG	2,500	.134
C	CpG	7,730	.414
G	GpG	4,960	.265
		18,680	1.000

Experiment IV with dCTP³²

Monomer Isolated	Doublet Sequence	Cpm	Fraction
T	TpC	4,130	.182
A	ApC	4,300	.189
C	CpC	6,070	.268
G	GpC	8,200	.361
		22,700	1.000

[a] From Josse *et al.* (1961).

tube, from a piece of naturally occurring DNA as "primer" and the four deoxynucleoside triphosphates. This was the basic experimental system of the nearest neighbor experiment. However, this time the objective of the experiment was to measure in a limited way the base sequence of the DNA synthesized. In order to do this, four separate experiments were performed for each kind of DNA. In each experiment a different one of the four nucleoside triphosphates was labeled with P^{32} in the phosphorus atom esterified to the 5′ carbon atom of the deoxyribose ring. The 5′ labeled kind of nucleotide was then incorporated into the DNA along with the other three unlabeled nucleotides. The DNA was hydrolyzed by the successive action of micrococcal DNase and calf spleen phosphodiesterase, which transfers the P^{32} to the 3′ position of the next sugar ring and effects a virtually complete breakdown of the DNA into 3′ nucleotides. Each of these nucleotides was then assayed for the amount of radioactivity it contained, and this amount was expressed as a fraction of the total radioactivity. Table 2 shows these radioactivity measurements for *M. phlei* DNA.

Take, for example, reaction I in which dATP³² carried the 5' label. When thymine was isolated as the 3' nucleotide, it was found to contain .075 of the total P³². The sum of these fractional values is 1 for each experiment, and each represents a relative frequency of occurrence out of a large number of possible occurrences because the DNA chain is very long. The fraction .075 does not represent simply the probability of occurrence of thymine in the DNA chain. For one thing, it is different in each of the four experiments. It represents only the thymine which occurred next to the incorporated radioactive adenine. It is the *conditional* probability of thymine in the 3' position, given that adenine occurred next to it in the 5' position. We may symbolize this as

$$3' \rightarrow 5'$$
$$.075 = p(\text{T} \mid \text{A})$$

It is important to understand that this is the internal direction of the linkage of the phosphate group from the 3' carbon atom of one pentose ring to the 5' carbon of the next in the interior of the chain. It is customarily denoted by a small p between the bases. However, the $3' \rightarrow 5'$ *internal* direction is the same as the $5' \rightarrow 3'$ *external* direction where, in this case, the number 5' denotes the free 5' terminal hydroxyl group and the 3' denotes the free 3' terminal hydroxyl group. These terminal hydroxyl groups may or may not be phosphorylated, but generally the 5' external hydroxyl group is phosphorylated and the 3' external hydroxyl is not. This is illustrated in the accompanying diagram and

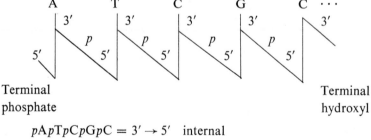

$$p\text{A}p\text{T}p\text{C}p\text{G}p\text{C} = 3' \rightarrow 5' \quad \text{internal}$$
$$= 5' \rightarrow 3' \quad \text{external}$$

equation. The literature is sometimes ambiguous, not clearly specifying whether the numbers, 3' and 5', are internal or external. We will always use the arrow to mean $3' \rightarrow 5'$ *internal*. This is extremely vital to the computations because the nearest neighbor frequencies are vectors, *not* scalars.

Table 2 lists the complete set of all 16 conditional probabilities for *M. phlei* DNA. Note that

$$p(\overrightarrow{T \mid A}) \neq p(\overrightarrow{A \mid T})$$
$$p(\overrightarrow{C \mid A}) \neq p(\overrightarrow{A \mid C}), \quad \text{etc.} \tag{25}$$

In the original experimental report (Josse *et al.*, 1961) these numbers were called "fractional base values," and it was not pointed out explicitly that they are conditional probabilities. I published this observation in 1963 (Gatlin, 1963). These conditional probabilities and the base composition were then inserted into equation set (23), which is just the definition of dependent events, and the doublet probabilities were calculated. They were called "nearest neighbor frequencies" and the experiment bears this name.

Kornberg did not measure the base composition of the DNA chemically. He calculated it from the conditional probabilities. This calculation utilizes some very fundamental summation properties of DNA matrices which we should examine in detail. Let us write the doublet probabilities as the matrix

5′	A	T	C	G
3′				
A	AA	AT	AC	AG] A
T	TA	TT	TC	TG
C	CA	CT	CC	CG
G	G̲A̲	GT	GC	GG
	A			

Any row or column sums to the probability of the single base common to the four doublets. We can prove that this must be the case. Let us write out the first column sum, expressing each doublet probability from equation set (23).

$$p(A)p(A \mid A) + p(A)p(T \mid A) + p(A)p(C \mid A)$$
$$+ p(A)p(G \mid A) = \; ?$$

We may factor out the $p(A)$:

$$p(A)[p(A \mid A) + p(T \mid A) + p(C \mid A) + p(G \mid A)] = ?$$

But the quantity in brackets is the sum of all the conditional probabilities for a single experiment and hence must sum to 1 (see Table 2). Therefore the column sum is simply

$$p(A)1 = p(A)$$

We may construct a proof for the row sums by a somewhat different procedure. The general conclusion is that, whenever we sum the probabilities of four doublets which have the same base at either the first or second position, this sum is equal to the probability of the common base. We may write the row sums as

$$p(A)p(A \mid A) + p(T)p(A \mid T) + p(C)p(A \mid C)$$
$$+ p(G)p(A \mid G) = p(A)$$
$$p(A)p(T \mid A) + p(T)p(T \mid T) + p(C)p(T \mid C)$$
$$+ p(G)p(T \mid G) = p(T)$$
$$p(A)p(C \mid A) + p(T)p(C \mid T) + p(C)p(C \mid C)$$
$$+ p(G)p(C \mid G) = p(C)$$
$$p(A)p(G \mid A) + p(T)p(G \mid T) + p(C)p(G \mid C)$$
$$+ p(G)p(G \mid G) = p(G)$$

$$(26)$$

Since all the conditional probabilities are known experimentally, we have four equations in four unknowns from which the base composition can be calculated. We will later describe notation which will enable us to write equation sets like (26) in a condensed form. The summation properties of these matrices above are fundamental to the development of the entropy H_{M}.

The conditional probabilities of the nearest neighbor experiment represent the first experimental measurement which can tell us something about the informational properties of the total base sequence of any large DNA molecule. It is a classic and historic experiment, comparable in importance to the early measurements of the spectrum of the hydrogen atom, which aided the development of quantum theory. One would think that, with the measurement of such an important property of DNA now possible, the literature would by now be flooded with nearest neighbor data. One would think that the experimentalist would want to characterize the DNA or RNA he has isolated by its matrix of conditional probabilities from which we can calculate its entropy, a property perhaps more important than its energy because of its relation to the informational capacity of the molecule. But the nearest neighbor experiment rests on the shelf, largely forgotten and unappreciated in the information explosion that is rocking the biological world today. A notable exception is the work of Subak-Sharpe and his collaborators (1966, 1969a, b).

3 *It takes two of us to discover truth:*
one to utter it and one to understand it.
—Kahlil Gibran, Sand and Foam

INFORMATION

DIMENSIONALITY

Almost always when we speak of transmitting information, we mean the transfer of some form of knowledge, factual or conceptual, by means of a linear sequence of symbols ordered according to the constraints of a language such as we are using here. We will confine our attention exclusively to the one-dimensional case for several good reasons. First, the obvious reason for considering only the one-dimensional case is that DNA itself stores its information in a one-dimensional sequence of symbols. Second, we can deal with it mathematically in a complete and satisfactory way.

But, apart from such obvious and practical reasons, there are interesting philosophical considerations. A linear sequence of symbols is the indispensable vehicle of communcation between higher organisms. Of course, two- or three-dimensional objects can convey information and in some instances are superior to language; but they are limited as a vehicle in a way that the one-dimensional vehicle is not. Any two- or three-dimensional object and, in fact, any general situation can be described by language and thus reduced to the one-dimensional case. This is precisely what a creative writer does and, if the description is reasonably objective, all observers can agree. For example, even an ink blot, although capable of eliciting detailed imaginative descriptions indicative of the underlying psychological makeup of the individual, can still be described objectively as just an "ink blot," and no one would disagree. Thus one can argue that any higher-dimensional object conveying information can always be reduced to the one-dimensional case, and at least some aspect of the information conserved.

The converse of this is not true if we include the concept of agreement mentioned above. It is not possible to take *any* arbitrary sequence of symbols and transform it into a three-dimensional form that will conserve even some part of the information. Although I might obtain an argumentative response to this proposition from an artist, it is certain that there would be no general agreement as to what the two- or the three-dimensional form should be. For example, what would be the three-dimensional representative of the sentence we are now reading? I see no way to construct rules for this transformation that all observers would agree upon. The element of reliability is lacking in the transformation and, as we shall discuss presently, this is a vital part of the capacity to transmit meaning.

Thus I regard the one-dimensional case as the most fundamental and, in a very important sense, the most general in information theory. The entropy concept arose from the three-dimensional thermodynamic setting and has evolved to its role in the description of linear sequences of symbols in information theory. This is precisely the opposite of the development of most concepts in mathematics and physics where one begins with the one-dimensional case, which is the simplest, and then generalizes to the n-dimensional case.

POTENTIAL VERSUS STORED INFORMATION

Recall that historically the development of information theory received its main impetus from the needs of electrical engineers working with communications problems. The communications engineer wishes to design in the most efficient manner possible a system which will transmit any message that may arise. He is interested in the *capacity to transmit* information, which I propose to call "potential information." As Weaver (1949) puts it, ". . . this word 'information' in communication theory relates not so much to what you do say, as to what you could say." Potential message variety, freedom of choice, and large vocabulary are the desired quantities to the communications engineer; and therefore he has come to associate the word information with this particular conceptual element. Since it varies directly with the entropy, high entropy means high potential information. In fact, the entropy of Shannon's formula, equation (15), is usually referred to as "information" throughout the literature and sometimes as "information content." The words "informational uncertainty" would be more accurate.

Potential information is vital to communication. Without this element of potential variety and uncertainty about what will come next, there can be no transmission of information. Cherry (1957) uses the illustration of a bookbinder's error that has made every page of a book the same. Only the first page could convey information. He states, "To set up communication, the signals must have at least some surprise value, some degree of unexpectedness, or it is a waste of time to transmit them." Thus with the higher entropy of potential information we associate the concepts of potential message variety, large vocabulary, surprisal value, and unexpectedness.

Are we to conclude then that as the entropy increases the information always increases? No, it is not quite this simple. The information concept is far richer than this. When I began to consider the problem of how much information is stored in the base sequence of a particular DNA molecule, I was faced with the following dilemma. We cannot simply equate high entropy with high information as the communications engineer has done. Let us take a simple example. A library obviously contains stored information The information is stored in a linear sequence of symbols ordered according to the constraints of a language. The sequences are organized into books and periodicals, and these are carefully ordered on shelves and neatly catalogued. Everywhere order and constraint are associated with the information storage process. This is a state of lowered entropy. If we were to take each page of each book, cut it into single-letter pieces and mix them in one jumbled heap, the entropy would unquestionably increase, but the stored information would decrease. Stored information, the "what we do say" of Weaver's statement, is associated with the ordering process brought about by the constraints of a language or any organized information storage process. Since stored information varies inversely with entropy, lowered entropy means a higher capacity to store information. This is precisely what we have been calculating in D_1 and D_2, whose sum measures exactly how much the entropy has been lowered from the maximum entropy state. Thus we may define stored information, I_s, as

$$I_s = D_1 + D_2 \qquad (27)$$

This distance on the entropy scale, $D_1 + D_2$, may be referred to by a number of names. I have just called it the "stored information." I have also referred to it as the "information content" of DNA (Gatlin, 1966, 1968). Another possible name is the "information density," which we will denote as I_d. This name has the advantage that it immediately implies a quantity which is independent of the *amount* of DNA in a cell. Also we shall show that, when this distance, $D_1 + D_2$, is divided by log a, it is equal to the quantity which Shannon called the "redundancy." Words tend to proliferate as does mathematical notation, but the satisfying thing about mathematical quantities themselves is their invariance.

LANGUAGE

In any information storage or transfer process the two mutually antagonistic forms of potential and stored information must seek an optimum; and nowhere is this dynamic optimization process more vividly illustrated than in the construction of language in general. It is well known that in any language the single-letter frequencies diverge from equiprobability as do the letter pairs, triplets, etc. A certain amount of ordering is necessary for the formulation of words and sentences. If there were no constraints and every possible letter combination occurred with random frequency, potential message variety would be maximal; but there would be no way to detect error because error detection and correction are based on forbidden and restricted combinations. For example, if we were confronted with the letter sequence thx, we would immediately reject it as a possible English word because it violates the rule or constraint that every English word must contain at least one vowel or y. More complex constraints play a role in the buildup of grammar and syntax. Any time we find these rules violated, we immediately suspect an error. But, if there were no rules at all, it would be impossible to tell if an error had occurred because in essence no error would have been defined. Thus it is possible for the potential information and entropy to be so high that transmission error makes communication impossible.

On the other hand, we can reduce the entropy to the point where the stored information becomes maximal, transmission is highly reliable; but the message variety is so low that we cannot say anything. In the extreme case we would have a monotone, a sequence of only one letter where the element of surprisal is completely lost.

Therefore, when formulating language, generally we wish to reduce the entropy somewhat but not too much, because we need an optimum blend of potential and stored information for successful or meaningful communication. We cannot store information without some message variety, and we cannot transmit without some reliability. Information is a capacity and entropy is its measure. However, the capacity to convey meaning through language depends not on an entropy maximum or minimum but rather on a delicate optimization of the two opposing elements of variety and reliability. If we carry either one to the extreme, we lose the meaning.

The reader will note that throughout this discussion I have not attempted to avoid the common usage of the word information as a synonym for meaning or knowledge. This usage will never be erased from our language. However, the phrases potential information and stored information, along with all their synonyms, are intended to refer to a *capacity* for communication, as separate and distinct from what is actually communicated.

Words such as variety, uncertainty, reliability, and fidelity clearly do not refer to the meaning of a message per se but to its *capacity* to convey this meaning. It is quite possible for a sequence of symbols that is complete gibberish in any human language to have the same informational *capacity* as a very meaningful sentence. Whenever we calculate the Shannon entropy of a sequence of symbols, it is the *capacity to transmit* or the potential information we measure. When we express this entropy as a divergence from the maximum value, $\log a$, this is a measure of the *capacity to store* or, more simply, the stored information. Most significantly for the living system, stored information is also a *capacity to combat error*.

We have spent considerable time discussing the conceptual aspects of information theory. This is important because information theory is a branch of mathematical knowledge, and all of mathematics is composed of two parts, the conceptual and the manipulative. Mathematics is simply a way of expressing concepts that anyone can understand in a way that very few can understand. The esoteric language is necessary, however, because it places at our disposal a powerful machinery which in turn gives rise to new concepts that would not have arisen without this manipulative process. We now turn to the manipulative aspects of information theory.

A MORE POWERFUL NOTATION

We have written S_1 as

$$S_1 = \{A, T, C, G\}$$

This is satisfactory when there are only four letters in the alphabet. But what if we had a very large number of letters? A more general notation would be to let x_i stand for any letter and let the index i take on the successive values from 1 to a, the number of letters in the alphabet. For example, for DNA we would let $x_1 = A$, $x_2 = T$, $x_3 = C$, $x_4 = G$. We could then write S_1 as

$$S_1 = \{x_i : i = 1, a\} \tag{28}$$

Instead of writing S_2 out as we did in equation (20), we could write

$$S_2 = \{x_i x_j : i, j = 1, a\} \tag{29}$$

This means that the indices i and j take on all possible values from 1 to a. It is convenient (although not absolutely necessary) to think of holding one index constant while the other runs over its total range of values as shown:

$$
\begin{array}{cccc}
x_1 x_1 & x_2 x_1 & x_3 x_1 & x_4 x_1 \\
\downarrow \begin{array}{c} x_2 \\ x_3 \\ x_4 \end{array} &
\downarrow \begin{array}{c} x_2 \\ x_3 \\ x_4 \end{array} &
\downarrow \begin{array}{c} x_2 \\ x_3 \\ x_4 \end{array} &
\downarrow \begin{array}{c} x_2 \\ x_3 \\ x_4 \end{array}
\end{array}
$$

We have thus represented all sixteen possible doublets with the notation of equation (29) and could easily represent a large number depending on the value of a. We will let p_i be the probability of x_i and p_{ij} be the conditional probability of x_j given x_i or, more simply, the probability of transition from x_i to x_j.

Let us note at this point some fundamental summations. The probabilities of all the single letters must always sum to 1:

$$\sum_i p_i = 1 \tag{30}$$

If the letters are independent events, the probability of any doublet sequence is $p_i p_j$, and these must also sum to 1:

$$\sum_i \sum_j p_i p_j = 1 \tag{31}$$

If the letters are not independent events, we must write the doublet probability as $p_i p_{ij}$, and again they sum to 1:

$$\sum_i \sum_j p_i p_{ij} = 1 \tag{32}$$

The reader should review at this point the discussion of the nearest neighbor experiment in Chapter 2. The nonindependent doublet frequency, $p_i p_{ij}$, is the "nearest neighbor frequency." The conditional probabilities or transition probabilities, p_{ij}, are the "fractional base values." The p_i are, as always, the "base composition" of DNA. The p_{ij} sum to 1 when the index, i, is held constant and we sum on the j,

$$\sum_j p_{ij} = 1 \tag{33}$$

but sum to nothing in particular when the j is held constant and we sum on the i. The "given" index of a conditional probability must be held constant to give a meaningful sum.

If for the doublet probabilities we hold one index constant and sum on the other, we obtain the probability of the single letter corresponding to the constant index. We proved this for one special case in Chapter 2. It makes no difference whether the doublet probabilities result from independent or nonindependent single letters. Thus

$$\begin{aligned}
\sum_i p_i p_j = p_j, \qquad \sum_i p_i p_{ij} = p_j \\
\sum_j p_i p_j = p_i, \qquad \sum_j p_i p_{ij} = p_i
\end{aligned} \tag{34}$$

Now let us write the general expression for H_2^{Ind}, the entropy of S_2 when the single letters are independent events:

$$H_2^{\text{Ind}} = -\sum_i \sum_j p_i p_j \log p_i p_j \tag{35}$$

and for H_2^{D}, the entropy of S_2 when the single letters are not independent events:

$$H_2^{\text{D}} = -\sum_i \sum_j p_i p_{ij} \log p_i p_{ij} \tag{36}$$

H_{M} IS AN ENTROPY

The last term in equation (35) is the log of the product $p_i p_j$, which may be broken into $\log p_i + \log p_j$. The complete split expression is

$$H_2^{\mathrm{Ind}} = -\sum_i \sum_j p_i p_j \log p_i - \sum_i \sum_j p_i p_j \log p_j \tag{37}$$

Since each symbol bears only one index, we may rearrange the summation signs thus:

$$H_2^{\mathrm{Ind}} = -\sum_i p_i \log p_i \sum_j p_j - \sum_j p_j \log p_j \sum_i p_i \tag{38}$$

The summation indices, i and j, are called "dummy" indices because they merely tell us to perform a sum. Hence they are quite arbitrary and *any* symbol can be used for this purpose. Thus $\sum_i p_i \log p_i$ is mathematically and conceptually identical to $\sum_j p_j \log p_j$. Therefore, since $\sum_i p_i = 1$ (equation 30),

$$H_2^{\mathrm{Ind}} = 2H_1 \tag{39}$$

Similarly we may take the expression for H_2^{D} apart:

$$H_2^{\mathrm{D}} = -\sum_i \sum_j p_i p_{ij} \log p_i - \sum_i \sum_j p_i p_{ij} \log p_{ij} \tag{40}$$

Here the quantity p_{ij} must be summed over both indices. If in the first term we hold the i constant and sum over the j, we have the sum

$$\sum_j p_i p_{ij} = p_i \tag{34}$$

This is the sum of all the doublet probabilities with a single base, x_i, common to all of them. We proved that this sum is the probability of the common base, in this case, p_i. Therefore

$$H_2^{\mathrm{D}} = -\sum_i p_i \log p_i - \sum_i \sum_j p_i p_{ij} \log p_{ij} \tag{41}$$

or

$$H_2^{\mathrm{D}} = H_1 + H_{\mathrm{M}} \tag{42}$$

where

$$H_{\mathrm{M}} = -\sum_i \sum_j p_i p_{ij} \log p_{ij} \tag{43}$$

The last term of equation (42) cannot be reduced further, and we have simply given it the name H_{M}, or "H-Markov." H_{M} must be an entropy because H_1 and H_2^{D} are both entropies and we cannot add something to an entropy and obtain another entropy unless the added quantity was itself an entropy also. It is now easy to show that H_M is the lower boundary of D_2 on the entropy scale (Fig. 5). We have defined D_2 as

$$D_2 = H_2^{\text{Ind}} - H_2^{\text{D}} \tag{24}$$

From equations (39) and (42),

$$D_2 = 2H_1 - H_1 - H_M \tag{44}$$
$$D_2 = H_1 - H_M \tag{45}$$

Since the divergence from independence cannot be negative, H_M can never be greater than H_1, i.e.,

$$H_M \leqslant H_1 \tag{46}$$

This inequality can be rigorously proved (Khinchin, 1957) and is sometimes referred to as Shannon's fundamental inequality.

ERGODIC MARKOV SOURCES

In order to define an ergodic Markov source, we must begin with some simpler definitions. In Chapter 2 we defined a random phenomenon. A *stochastic process* is a random phenomenon which follows a time course of development controlled by probabilistic laws. One may make observations on the outcomes of a stochastic process at various points along a time axis. A *Markov process* is a stochastic process wherein the physical system undergoing the process is characterized by a set of discrete "states" which correspond to the different possible outcomes of the random phenomenon. There is a definite probability of transition from one state to the next. If the conditional probability, p_{ij}, that the system will be in state j at time t, given that it was in state i at time $t-1$, does not depend on the state of the system *before* $t-1$, then the process is called a *Markov* process, and the sequence of events or *states* which, of course, may be represented by a sequence of symbols is called a *Markov chain*. This definition implies that $p_{ijk} = p_{jk}$. In words, the conditional probability that the system will be in state k at time t, given that it was in state j at time $t-1$ and in state i at $t-2$ is the same as the simple conditional probability of k given j, because the fact that the system was in state i before this has no influence. One sometimes says that a Markov chain has no "memory" of its past.

This is the classical definition of a Markov chain. However, upon introduction of the source concept, information theorists have taken some liberties in interpreting the meaning of the "state" of a system undergoing a stochastic process.

A *source* is simply any device or process which emits a sequence of symbols. The source may have a physical representation such as a radio transmitter, but this is not necessary. One may select a set of sequences and choose to regard them as the output of a "source." It is useful to consider the statistical properties of the source in general rather than any particular one of the sequences which it may emit.

An information source is characterized by a finite set of symbols, called the source alphabet, which it can emit. If we were to regard each source symbol as a "state" of the system, not all information sources would be Markov sources under the classical definition because the probability of transition from one symbol to another may depend not only on the immediately preceding symbol but also on the m preceding symbols where $m = 0, 1, 2, 3$, etc. The simplest kind of information source where $m = 0$ is called a *zero-memory* source and the symbols which it emits are all statistically independent of each other. When $m = 1$ we have a *first-order Markov source*. When $m = 2$ we have a *second-order Markov source*, etc. When $m = 2$, the probability of emission of a particular symbol depends on the two preceding symbols, i.e.,

$$p_{ijk} \neq p_{jk} \quad \text{for all} \quad i, j, \text{and } k$$

m is called the memory of the information source, and the m preceding symbols are regarded as the "state" of the mth-order Markov source at the previous time.

Mathematicians recognize this possibility of defining blocks of symbols rather than single symbols as the previous "state" of a Markov chain. Parzen (1962) defines a stochastic process, called a "multiple Markov chain," and discusses its application to human language where the letter dependencies clearly reach farther back than one letter.

m is finite and the Markov source has no "memory" of its history preceding the m-symbol block. In practice, the dependence on the previous symbol may decrease continuously such that the cutoff point defining m is somewhat arbitrary and perhaps dependent on observational limitations. In such cases we may regard the mth-order Markov sources of a series of sources of increasing m as successive approximations to the "true" Markov source. Shannon's original paper (Shannon, 1949) describes a series of information sources of increasing memory which approximate the English language closer and closer as m increases.

It is clear from the discussion above that up to this point we have treated DNA as the output of a first-order Markov source. If higher-order dependencies exist, then this is a first approximation. Higher-order effects probably exist. However, at present we have no experimental data on the higher-order statistical properties of *long* nucleic acid sequences comparable in scope and generality to the nearest neighbor data. The nearest neighbor experiment, as its name so aptly describes, measures only the nearest neighbor or first-order effect. Therefore, when we apply our mathematical development to the presently existing nearest neighbor data, we are limited to the calculation of a first-order dependence.

The nearest neighbor experiment measures experimentally the first-order transition probability matrix, the sixteen p_{ij}. In all DNAs or RNAs examined so far the conditional probabilities are not identical with the base composition values, i.e.,

$$p_{ij} \neq p_j$$

for the entire set of transition probabilities. This in itself shows that DNA is a linear sequence of symbols where the probability of occurrence of a single base in the chain is dependent on the base which immediately preceded it in the chain. This means that DNA is the output of *at least* a first-order Markov source.

We noted at the outset of our discussion that one may select a particular set of sequences and regard them as the output of a single source. We must now examine the criteria which would make this selection process realistic. We need more definitions.

A *stationary source* is one whose probabilistic laws are time-invariant. There is a specific type of stationary source called an *ergodic* source whose mathematical definition is beyond the scope of this book. The concept is subtle, and different authors stress different aspects of the concept. Abramson (1963) states:

> An ergodic source is merely a source which, if observed for a very long time, will (with probability 1) emit a sequence of source symbols which is "typical." In fact, the idea of sources possessing the ergodic property is so natural that some readers may have difficulty in picturing a source which is *not* ergodic.

Abramson then gives an example of a binary source which, if it ever reaches the state 00, remains "stuck" in that state and all succeeding symbols are zeros.

Thus for a nonergodic source the sequence which one would observe is highly dependent on the initial point of the observation. This is not the case with ergodic sources. It does not matter what initial state we choose when observing the transition probability distribution of an ergodic source. For a sufficiently long sequence (or period of time) the *m*-step transition probabilities all approach stable limiting values which are independent of the initial state and are "typical" or characteristic of the source. This unique probability distribution is called the *stationary distribution* of the ergodic Markov source.

Let us apply the ergodic source concept to DNA. First we must define the source. We could choose to regard the DNA of all living organisms as the output of a single massive evolutionary source, but this source would not be ergodic. It is not even stationary.

Here we see that the concept of stationarity is relative to the definition of a time scale. If a source is evolving, as living sources are, it may be regarded as stationary over a relatively short time period but as nonstationary on a larger evolutionary time scale.

We could choose to regard the DNA of a particular species as the output of a source. For many lower organisms we might reasonably regard this source as ergodic. For example, we could choose to regard the base sequence of the DNA of all *E. coli* organisms as the output of a single ergodic first-order Markov source. This choice is an assumption and an approximation subject to the following limitations.

In the nearest neighbor experiment the "primer" DNA is a sample of the DNA of the organism or tissue under observation. It is clear that this sample must be representative of the entire population of DNA molecules in the classical statistical sense. In fact this concept is sometimes used to define the ergodic concept, as the following quotation from Parzen (1962) illustrates.

> In general, a stochastic process is said to be ergodic if it has the property that the sample (or time) averages formed from an observed record of the process may be used as an approximation to the corresponding ensemble (or population) averages.

Therefore, if we wish to regard the DNA of a particular organism or group of organisms as the output of an ergodic source, the primer DNA of any organism tested must be representative of the population, within the experimental error of the method. The nearest neighbor frequencies (and possible future m-step transition probabilities) must be stable, reproducible, and characteristic of that particular tissue, organism, or group of organisms. This appears to be the case for many of the DNAs examined so far, particularly for lower organisms. In fact, even a relatively small percentage (20%) of the primer DNA is representative of the total DNA (Swartz et al., 1962).

However, the relatively recent observations of satellite components of the DNA of higher organisms (above bacteria) complicate the ergodic assumption. When we centrifuge the DNA from a higher organism in a density gradient, we observe a "main" peak and sometimes one or more smaller satellite peaks, indicating inhomogeneity of the population of DNA molecules under a density criterion. The DNA of the crab species *Cancer borealis* consists of two components under density centrifugation, a main component and a satellite component comprising about 30% of the total DNA. The nearest neighbor frequencies for both the main DNA and the satellite have been measured. They are distinctively different. However, each different probability distribution is stable, reproducible, and characteristic of its corresponding component DNA. It is clear that in such a case we could regard the crab DNA as the output of a "composite" source with two component sources, each of which could be regarded as an ergodic source, the accuracy of this assumption being directly proportional to the resolution of the component peaks. We are essentially saying that the physical representatives of sufficiently long output sequences from an ergodic source should form a unimodal distribution with a low variance under any criterion designed to test for inhomogeneity.

In Chapter 7 we shall discuss another criterion of inhomogeneity in the DNA of higher organisms, the rate of renaturation of dissociated DNA strands. If one plots the amount of DNA versus the rate of re-association, certain distinct classes of DNA appear as "peaks" in a spectrum. In several cases observed so far, for example, the mouse satellite and the guinea pig satellite, the density gradient and re-naturation criteria coincide, i.e., the DNA component which re-associates at a different rate from the main DNA also has a different density. When we observe a component of the DNA of higher organisms that is distinguishable by independent inhomogeneity criteria and that displays a characteristic, stable, reproducible transition probability matrix, it is reasonable to assume that this DNA may be regarded as the output of an ergodic Markov source.

We are essentially assuming this whenever we report the nearest neighbor frequencies of any DNA. If the source were not stationary, this measurement would vary with time. If the source were not ergodic, different primer DNA samples might give different nearest neighbor frequencies. When we calculate quantities such as D_2, which according to the definition in equation (21a) of the next section contain limiting values, in order that the computation be meaningful the function must converge and the limits must exist. In particular, it can be shown that

$$\lim_{n \to \infty} \frac{H_n^D}{n}$$

exists for all ergodic sources (Khinchin, 1957).

As our discussion above shows, the concepts of stationarity and ergodicity are not absolutes. We may say that a source is "reasonably" stationary and ergodic relative to our definitions and criteria. The most practical way to show that one's definition of ergodicity is realistic is simply to calculate H_n^D/n as a function of n and see if it exhibits an asymptotic behavior. From equation (12a) of the next section, it is a simple algebraic exercise to show that

$$\frac{H_n^D}{n} \geqslant \frac{H_{n+1}^D}{n+1} \tag{47}$$

FIGURE 6

H_n^{D}/n VERSUS n FOR RABBIT LIVER DNA.
THE DASHED LINE IS H_{M}.

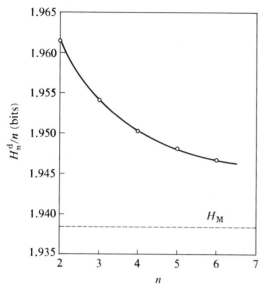

From Gatlin (1968).

Thus a plot of H_n^{D}/n should approach H_{M} as a *lower* limit. All the experimental data for DNA verify this prediction. Figure 6 shows a typical plot. These calculations could be extended to higher n and the asymptotic approach would look even better, but they are somewhat laborious.

An important matter in defining an ergodic source is the limitation of inhomogeneity criteria. In defining composite sources, one must not make the inhomogeneity criteria too stringent. For example, the rapidly renaturing DNA discovered by Britten and Kohne (1967) is believed to be interspersed throughout the main DNA which reassociates several orders of magnitude more slowly. Even such composite DNA may be regarded as the output of an ergodic source if these rapidly reassociating sequences (which I believe to be highly redundant control sequences) are scattered uniformly throughout the composite sequence.

The situation is closely analogous to the stationarity problem discussed by Neyman and Scott (1957). Parzen (1962) comments on this discussion as follows:

It should be noted that the fact that a stochastic process is stationary does not mean that a typical sample function of the process has the same appearance at all points of time. Suppose that at some time in the history of the world an explosion takes place in some very isolated normally quiet forest. Then [the above process]·is a strictly stationary process if the time at which the explosion occurs is chosen uniformly on the real line.

Thus even the inhomogeneity criterion itself is subject to definition and limitation.

Throughout the preceding discussion we have been dealing with the subtle and intricate business of applying a mathematical concept to the real world. We have long recognized this to be an "art," and anyone who does it successfully is as much an artist as a Van Gogh or Picasso. I regard the ergodic concept as one of the most beautiful concepts in all of mathematics.

THE GENERAL MARKOV SOURCE

In this section I wish to present the mathematical formulation of my theory in a condensed, complete, and general form. This section is completely self-contained, including the equation numbers, and may be omitted without loss of vital information.

The reader with a mathematical background perhaps will have been somewhat dismayed at the lack of rigor and generality in the previous sections. I hope that this section will be at least partially redemptive and serve as a convenient summary.

We are dealing with language in general. We first define the alphabet:

$$S_1 = \{x_i : i = 1, a\} \tag{1a}$$

where the x_i are the letters of the alphabet and a is the number of letters. With each x_i there is associated a probability, $0 \leqslant p_i \leqslant 1$:

$$\sum_i p_i = 1 \tag{2a}$$

Thus S_1 is a finite probability space. The entropy of S_1 is

$$H_1 = -K \sum_i p_i \log p_i \tag{3a}$$

The maximum entropy state for a sequence of symbols is characterized by equiprobable, independent elementary events, x_i, on S_1. If the sequence diverges from the maximum entropy state, the divergence from equiprobability, D_1, is given by

$$D_1 \equiv H_1^{\text{Max}} - H_1^{\text{Obs}} \tag{4a}$$

where $H_1^{\text{Max}} = \log a$ and the p_i on S_1 are the experimentally observed values for a given language. Thus

$$D_1 = \log a - H_1 \tag{5a}$$

We are interested in the sequential arrangement of the letters in the sequence. Therefore we define a space of n-tuples:

$$S_n = \{x_i x_j \ldots x_n : i, j, \ldots, n = 1, a\} \tag{6a}$$

There are a^n n-tuples in S_n. If the letters in the n-tuple are independent of each other,

$$H_n^{\text{Ind}} = -\sum_i \sum_j \cdots \sum_n p_i p_j \cdots p_n \log p_i p_j \cdots p_n \tag{7a}$$

or

$$H_n^{\text{Ind}} = nH_1 \tag{8a}$$

Let m be the memory of a Markov source. If $m = 1$, the entropy of S_n is given by

$$H_n^{\text{D}} = -\sum_i \sum_j \cdots \sum_n p_i p_{ij} \cdots p_{(n-1)n} \log p_i p_{ij} \cdots p_{(n-1)n} \tag{9a}$$

where p_{ij} is the Markov transition probability from letter i to letter j. Utilizing the summations

$$\sum_j p_{ij} = 1 \tag{10a}$$

and

$$\sum_i p_i p_{ij} = p_j \tag{11a}$$

equation (9a) reduces to

$$H_n^{\text{D}} = H_1 + (n - 1)H_M^{(1)} \tag{12a}$$

where

$$H_M^{(1)} = -\sum_i \sum_j p_i p_{ij} \log p_{ij} \tag{13a}$$

The superscript on H_M serves to remind us of the memory of the source If $m = 2$,

$$H_n^D = -\sum_i \sum_j \cdots \sum_n p_i p_{ij} p_{ijk} \cdots p_{(n-2)(n-1)n}$$

$$\times \log p_i p_{ij} p_{ijk} \cdots p_{(n-2)(n-1)n} \quad (14a)$$

or

$$H_n^D = H_1 + H_M^{(1)} + (n - 2)H_M^{(2)} \quad (15a)$$

where

$$H_M^{(2)} = -\sum_i \sum_j \sum_k p_i p_{ij} p_{ijk} \log p_{ijk} \quad (16a)$$

Following this pattern, we generalize for the mth-order Markov source.

$$H_n^D = H_1 + H_M^{(1)} + H_M^{(2)} + \ldots H_M^{(m-1)} + (n - m)H_M^{(m)} \quad (17a)$$

where

$$H_M^{(m)} = -\sum_{i_\sigma=1}^a p_{i_1} p_{i_1 i_2} p_{i_1 i_2 i_3} \cdots p_{i_1 \ldots i_{(m+1)}} \log p_{i_1 \ldots i_{(m+1)}} \quad (18a)$$

Since

$$H_M^{(m+1)} \leqslant H_M^{(m)} \quad (19a)$$

(This is a generalized form of Shannon's fundamental inequality whose proof is given by Khinchin, 1957.)

$$H_n^{\text{Ind}} \geqslant H_n^D \quad (20a)$$

If the sequence of symbols diverges from the maximum entropy state owing only to a divergence from independence of the symbols, this divergence must be a function of $(H^{\text{Ind}} - H_n^D)$. Since both of these entropies are monotonically increasing functions of n, I define the divergence D_2 from independence as

$$D_2 \equiv \lim_{n \to \infty} \frac{1}{n} (H_n^{\text{Ind}} - H_n^D) \quad (21a)$$

From (8a),

$$\lim_{n \to \infty} \frac{H_n^{\text{Ind}}}{n} = H_1 \quad (22a)$$

and, from (17a) if we impose the condition that $m \ll n$,

$$\lim_{n \to \infty} \frac{H_n^D}{n} = H_M^{(m)} \quad (23a)$$

Therefore

$$D_2 = H_1 - H_M \tag{24a}$$

where the memory of the source is understood. If one knows m for any given language, one can always impose the condition $m \ll n$ simply by considering sequences of sufficient length. For DNA, $n \simeq 10^6$ to 10^9 whereas m is a small integer. Therefore $m \ll n$ holds for DNA.

The reader with background in information theory will note an analogy between our expression for D_2 and the classical quantity referred to by various names such as the mutual information of a channel, the transinformation, the rate of information transmission. The classical form is (after Abramson, 1963)

$$I(A; B) = \sum_{A,B} p(a, b) \log \frac{p(a \mid b)}{p(a)} \tag{25a}$$

which in our i, j notation is

$$I(I; J) = \sum_i \sum_j p_i p_{ij} \log \frac{p(i \mid j)}{p_i} \tag{26a}$$

However, the expression for our D_2 is

$$D_2 = \sum_i \sum_j p_i p_{ij} \log \frac{p(j \mid i)}{p_i} \tag{27a}$$

These forms can be shown to be equivalent for any directed sequence when one considers the interaction of the sequence direction vector and the direction in which the conditional probabilities are calculated, i.e., as "forward" or "backward" under Bayes' theorem.

Let us rewrite equation (13a) for $m = 1$ in the form

$$H_M = -\sum_{i,j} p(ij) \log p(j \mid i) \tag{28a}$$

The classical expression is always written

$$H_M = -\sum_{i,j} p(ij) \log p(i \mid j) \tag{29a}$$

For a directed sequence where i and j refer to the same alphabet, these forms are equivalent. Proof follows.

Let us denote the sequence direction vector as $3' \rightarrow 5'$. Equation (28a) becomes

$$H_M = -\sum_{i,j} p(\overset{3' \, 5'}{i \, j}) \log p(\overset{5' \quad 3'}{j \mid i}) \tag{30a}$$

But the i and j are dummy indices. Hence an equivalent expression is

$$H_M = -\sum_{i,j} p(\overset{3'\,5'}{j\,i}) \log p(\overset{5'\,3'}{i\,|\,j}) \tag{31a}$$

But

$$p(\overset{3'\,5'}{j\,i}) = p(\overset{5'\,3'}{i\,j}) \tag{32a}$$

so

$$H_M = -\sum_{i,j} p(\overset{5'\,3'}{i\,j}) \log p(\overset{5'\,3'}{i\,|\,j}) \tag{33a}$$

And since the sequence direction is arbitrary, we may reverse it (again). We have

$$H_M = -\sum_{i,j} p(\overset{3'\,5'}{i\,j}) \log p(\overset{3'\,5'}{i\,|\,j}) \tag{34a}$$

Hence we have shown that equations (30a) and (34a) are equivalent sums even though $p(\overset{3'\,5'}{i\,|\,j})$ and $p(\overset{5'\,3'}{j\,|\,i})$ are *not* equivalent quantities.

Perhaps we should note, for example, that

$$H_M = -\sum_{i,j} p(\overset{3'\,5'}{i\,j}) \log p(\overset{5'\,3'}{i\,|\,j}) \tag{35a}$$

is an inconsistently labeled expression and gives completely meaningless results in biological calculations.

This is an interesting situation. We are in essence dealing with two different types of directed quantities, the sequence direction vector and the "direction" of the conditional probability which we could call the "Bayes vector."

Thus H_M and hence D_2 are multicomponent quantities which involve the interaction of two different types of vectors and are symmetric with respect to reversal of either of these vectors or the interchange of their defining indices. This is a type of symmetry which is characteristic of certain fundamental tensors.

At this point let us analyze the structure of D_2. The most general form of equation (24a) is

$$D_2^{(m)} = H_1 - H_M^{(m)} \tag{36a}$$

H_M is a limiting value of H_n^D/n and has the structure shown in equation (17a). It is apparent from this structure that as we consider higher and higher orders of m, $D_2^{(m)}$ is increased in increments of

$$D_\lambda = H_M^{(\lambda-2)} - H_M^{(\lambda-1)} \tag{37a}$$

where λ is always one more than the highest memory of the source under consideration. For example, equation (24a) may be written

$$D_2^{(1)} = H_M^{(0)} - H_M^{(1)} \tag{38a}$$

where $H_M^{(0)}$ is equivalent notation for H_1. This is the first increment of $D_2^{(m)}$. The second increment is

$$D_3 = H_M^{(1)} - H_M^{(2)} \tag{39a}$$

where now

$$D_2^{(2)} = D_2^{(1)} + D_3 \tag{40a}$$

and in general

$$D_2^{(m)} = D_2^{(1)} + D_3 + D_4 + \ldots + D_{(m+1)} \tag{41a}$$

If we allow the range of m to include zero, we may think of $D_2^{(0)}$ as equivalent notation for D_1. Then the total divergence from the maximum entropy state (the information density) is given by

$$I_d = D_1 + D_2 + D_3 + \ldots + D_{(m+1)} \tag{42a}$$

The increments of $D_2^{(m)}$ are shown in Fig. 7.

I wish to acknowledge that Temple F. Smith and Thomas A. Reichert both suggested to me that there must be a D_3, D_4, etc.

FIGURE 7

DETAILED ENTROPY SCALE.

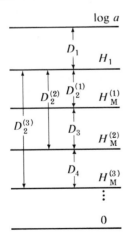

Although D_1 and D_2 were originally described as the divergence from equiprobability and the divergence from independence, respectively, *of the single letters* in the alphabet, it is also possible to describe all divergence from the maximum entropy state purely in terms of divergence from equiprobability *of the n-tuples*.

Thus $D_2^{(1)}$ is the divergence from equiprobability of the 2-tuples over and above that imposed upon the distribution by D_1. For example, for the DNA space, if the letter A is of very low frequency, this requires that the n-tuples containing large amounts of A, such as $AAAAA\ldots$, must be scarce. In general D_λ is the divergence from equiprobability of the λ-tuples over and above that imposed upon the distribution by the lower level divergences, $D_{(\lambda-1)}, D_{(\lambda-2)}, \ldots, D_1$.

Since all ergodic sources possess the E property (see Khinchin, 1957), there is a limit to this process of adding increments of $D_2^{(m)}$. The E property predicts that for sufficiently large m, a space will be reached where all the a^λ sequences of length λ become divided into two groups, the high probability group and the low probability group. The sum of the probabilities of all the sequences in the low probability group approaches zero and the probability of each sequence in the high probability group approaches $2^{-\lambda H_M(m)}$. Thus the λ-tuples all become equiprobable for sufficiently large λ if the source is ergodic.

Beyond this section we will return to our simpler notation where the memory of the source is understood to be 1 unless otherwise stated.

INFORMATION DENSITY OF A MARKOV CHAIN

H_M is the entropy of a first-order Markov chain. Note that the expression for H_M, equation (43), contains both the p_i and the p_{ij}. The distance of H_M from the maximum entropy value, log a, should contain both the divergence from equiprobability and the divergence from independence. This is easy to show. The total information density of a first-order Markov chain is

$$I_d = \log a - H_M \tag{48}$$

since

$$D_1 = \log a - H_1 \tag{19}$$

and

$$D_2 = H_1 - H_M \tag{45}$$

$$I_d = D_1 + D_2 \tag{49}$$

THE REDUNDANCY CONCEPT

Classical information theorists were not unaware of the concept of information density. In fact, Shannon defined a concept called the "redundancy" which is essentially the same thing. The ratio of the entropy of an information source, or the sequence of symbols it emits, to the maximum entropy value, $\log a$, is called the relative entropy. For DNA this ratio is $H_M/\log a$. Shannon defined the redundancy, R, of a sequence of symbols as one minus the relative entropy.

$$R = 1 - \frac{H_M}{\log a} \tag{50}$$

This may be written as

$$R = \frac{\log a - H_M}{\log a} \tag{51}$$

From (19) and (45),

$$R \log a = D_1 + D_2 \tag{52}$$

This is a fundamental equation. Let us note some of its properties. The constancy of D_1 implies that R versus D_2 is a straight line with slope $1/\log a$ and intercept $D_1/\log a$. Conversely, if R versus D_2 is observed to be linear with slope α and intercept ω,

$$R = \alpha D_2 + \omega \tag{53}$$

From (52) this implies that D_1 is a linear function of D_2 with slope $(\alpha \log a - 1)$ and intercept $\omega \log a$:

$$D_1 = (\alpha \log a - 1)D_2 + \omega \log a \tag{54}$$

The case when D_1 is a constant occurs when $(\alpha \log a - 1) = 0$ and $\alpha = 1/\log a$. Thus when R versus D_2 is linear, the coordinates (R, D_2) define a vector which we shall call the $RD2$ vector (or the $D2R$ vector is equally suitable).

Reichert and Wong (1971) were the first to observe the linearity of R versus D_2 on the protein space of cytochrome c. They remark that this line is "truly time's arrow." There are many elements of their techniques of observation which are now under evaluation, and we do not yet know the limitations of these techniques. But there can be no doubt that their "arrow," to which we shall often refer simply as "the vector," is a fundamental invariant, and they must be given full credit for its discovery. It is, in fact, the first observation of a true biological invariant, a tensor of rank one. We shall return to its description on the DNA space many times.

From (54), (19), and (45),

$$H_1 = \frac{(\alpha \log a - 1)}{\alpha \log a} H_{\mathrm{M}} + \frac{(1 - \omega)}{\alpha} \qquad (55)$$

Thus we have a hierarchy of linear relations when R versus D_2 is observed to be linear.

From (49) and (52),

$$R \log a = I_d \qquad (56)$$

or

$$R = \frac{I_d}{\log a} \qquad (57)$$

Shannon's redundancy *is* the information density expressed as a fraction of its maximum value, $\log a$. For a given language where the number of letters is constant, R and I_d are directly proportional to each other. We may think of them in a general way as the same fundamental quantity except for the scaling factor, $\log a$.

The word redundancy is again one of those words in information theory with a common usage and an assigned technical meaning. The common usage of the word refers to simple repetition, superfluous to meaning. This kind of redundancy we will call repetitive redundancy and, whenever it is necessary to avoid ambiguity, denote it by R-redundancy. The obvious purpose of repetition is to get a message across with a minimum of error in the presence of interfering noise. R-redundancy does help to combat error, but it is a costly kind of insurance.

We have already explained that information density is a capacity to combat error. It measures how much the entropy has been lowered from its maximum value and thus is a measure of all the ordering, constraints, rules, etc., that have been imposed upon the system. This kind of redundancy as defined by Shannon we will call S-redundancy. S-redundancy combats error, but in a completely different way from R-redundancy. S-redundancy is a direct measure of all the rules which define error in a language. The English language has such a high S-redundancy that we can sometimes strike out as many as 50%, perhaps sometimes even 75%, of the symbols in a sentence and still reconstruct the sentence. This means that about half of the symbols are a matter of free choice and the other half are fixed by the rules of the language. This kind of redundancy is a very effective safeguard against error.

The conventions S-redundancy and R-redundancy are somewhat cumbersome. We will use them whenever a situation arises that would cause confusion without them. However, their continued usage is usually unnecessary. From this point on, the word redundancy and the symbol R will refer exclusively to the Shannon redundancy unless otherwise stated.

REDUNDANCY **STRUCTURE**

$R \log a$ is the information density and is the sum of D_1 and D_2. D_1 and D_2 are subject to the following constraint. Since

$$H_M \leqslant H_1 \tag{46}$$

and

$$D_2 = H_1 - H_M \tag{45}$$

D_2 is limited to the range 0 to H_1. This is easily visualized by referring to the entropy scale, Fig. 5. Once D_1 is fixed, as it is for any DNA of a given base composition, then H_M may be equal to H_1, in which case $D_2 = 0$, or it may descend the scale to the ultimate value of zero. After D_1 is fixed, D_2 may have any value in this range.

Thus it is possible for the total divergence from the maximum entropy state to be composed of differing relative amounts of D_1 and D_2. For example, a given value of I_d may be reached with $D_1 \geqslant D_2$ or vice versa. In fact, the same value of I_d can be achieved in an infinite number of ways depending on the relative contributions of D_1 and D_2.

Therefore we define an index to characterize this relative contribution.

$$RD1 = \frac{D_1}{D_1 + D_2} = \frac{D_1}{I_d} = \frac{D_1}{R \log a} \tag{58}$$

$$RD2 = \frac{D_2}{D_1 + D_2} = \frac{D_2}{I_d} = \frac{D_2}{R \log a} \tag{59}$$

$$RD1 + RD2 = 1 \tag{60}$$

There is, of course, only one independent index being defined; but it is sometimes convenient to speak of the contribution of D_1 to R, $RD1$, or of D_2 to R, $RD2$. We will call these indices the divergence indices or simply the D-indices. In a previous publication (Gatlin, 1968), I labeled the quantity $RD2$ simply as B. I find the names $RD1$ and $RD2$ more descriptive.

We have arrived at a new concept via the manipulative machinery. If we were given two identical values of R, each with a significantly different D-index, we would have a sequence with the same amount but different *kinds* of redundancy. I_d or R tells us *how much* divergence there has been from the maximum entropy state, and the D-indices tell us *what kind* of divergence it is, i.e., whether it is composed mostly of D_1 or D_2. Since R and the D-indices are both fractions, they are dimensionless and have the value range 0 to 1.

We may describe this entire situation in another way. Since the coordinates (R, D_2) define a vector, the coordinates $(R, RD2)$ or $(R, RD1)$ are alternative descriptions of this same vector: given R and either D-index, D_2 can be calculated. We have certainly added another dimension to the classical entropy concept; and when one considers higher-memory Markov sources, higher dimensions appear. Referring to the notation of the self-enclosed mathematical section, one can define $RD3$, $RD4$, etc. If the memory of the Markov source is m, it takes $(m + 1)$ coordinates to specify completely the structure of the redundancy (or information density).

Since the nearest neighbor data limit us to the consideration of a first-order Markov dependence, we will consider only two coordinates throughout most of this book. Let us now observe how these two coordinates describe living systems and what they tell us about their evolution.

4

*I have kindled the dark depths
of beginning life and watched it crawl
from caves to rocky heights.*
—*Kahlil Gibran, The Earth Gods*

LIVING SYSTEMS

THE PRIMITIVE QUESTION

We are now in a position not only to answer the primitive question, "How much information is stored in a given DNA molecule?" but also to ask it in a more definitive manner. We must first ask, "What is the Shannon entropy of a given DNA molecule?" For DNA this must be approximated as the entropy of a first-order Markov chain, H_M. This number measures directly the potential information or the potential message variety of the sequence which classical information theorists have called simply "information." If we wish to answer the question "What is the capacity of the sequence to store information?" we must express this point on the entropy scale as a distance from the maximum, either as I_a, which is equal to $R \log a$, or simply as R, the redundancy. The question we are really asking is "What is the fidelity of the message?" since, as we have discussed, redundancy is a measure of all the constraints on a language which make error detection and correction possible. It is in this sense a direct measure of the reliability of a message. Hence a concise way to ask the primitive question is "What is the redundancy of a given DNA molecule?"

This formulation has the advantage of utilizing the familiar concept of redundancy from classical information theory. We may then ask, in addition, "What are the *D*-indices?" This question has no counterpart in classical theory. After we have answered both of these questions, we will know both the *amount* and *kind* of information density of the sequence in question. We are now ready to answer these questions quantitatively.

TABLE 3[a]

INFORMATIONAL PARAMETERS

Organism or Tissue	D_1	D_2	R	$RD2$
PHAGE				
T4	.0590	.0045	.0318	.0708
T2	.0533	.0057	.0295	.0962
T6	.0532	.0049	.0291	.0837
T5	.0470	.0003	.0237	.0065
ϕX–174*	.0116	.0125	.0121	.5188
T1*	.0070	.0115	.0093	.6212
λdg*	.0000	.0145	.0073	1.0000
$\lambda+$*	.0000	.0107	.0054	1.0000
BACTERIA				
M. lysodeikticus	.1286	.0132	.0709	.0928
M. phlei	.0892	.0273	.0582	.2342
H. influenza	.0405	.0225	.0315	.3574
B. subtilis*	.0097	.0200	.0149	.6725
A. aerogenes*	.0094	.0196	.0145	.6757
E. coli B$_b$*	.0017	.0200	.0109	.9221
E. coli B$_a$*	.0000	.0159	.0080	1.0000
PROTOZOA				
Tetrahymena pyriformis	.2115	.0138	.1127	.0615
Chlamydomonas*	.0052	.0197	.0125	.7916
PLANT				
Wheat germ*	.0078	.0094	.0086	.5466
INVERTEBRATE				
Crab testis (light comp.)	.8332	.8003	.8168	.4900
Crab testis (main comp.)	.0600	.0112	.0356	.1568
Paracentrotus lividus	.0586	.0072	.0329	.1092
Echinus esculenta	.0484	.0153	.0319	.2408
Starfish testis	.0280	.0162	.0221	.3670
VERTEBRATE				
Mouse ascites tumor	.0232	.0436	.0334	.6520
Chicken red cell	.0177	.0467	.0322	.7250
Human spleen	.0256	.0380	.0318	.5975
Mouse liver	.0210	.0409	.0309	.6606
Mouse lymphoma	.0192	.0419	.0305	.6852
Rabbit liver	.0169	.0435	.0302	.7207
Mouse thymus	.0195	.0367	.0281	.6533
Bovine sperm	.0171	.0381	.0276	.6904
Calf thymus	.0138	.0394	.0266	.7405
Bovine thymus	.0108	.0424	.0266	.7967
Bovine liver	.0107	.0394	.0251	.7858
Salmon liver	.0147	.0290	.0219	.6631

[a] All nearest neighbor data are from Josse et al. (1961) and Swartz et al. (1962). Table is from Gatlin (1968).

* This falls on the $RD2$ vector.

LIMITATIONS OF THE DATA

The two fundamental quantities which one calculates are H_1 and H_M. From these one can calculate D_1, D_2, R or I_d, and $RD1$ or $RD2$. I have chosen to list D_1, D_2, R, and $RD2$. From these and the relations in Chapter 3, one can calculate any of the other quantities.

Table 3 lists D_1, D_2, R, and $RD2$ for the nearest neighbor data of 35 organisms and tissues. Table 4 shows a comparison of DNA and the RNA synthesized from the same DNA as a template. The values are very close, in many cases, within experimental error.

Technically, the last digit listed is not significant. However, every quantity with which we deal involves ultimately H_1 or H_M or both, which are sums over all states; hence the random experimental error will tend to cancel out to some extent. The situation is analogous to the integration of a weak spectroscopic signal in the presence of background noise. The noise is random; hence there are roughly as many positive fluctuations as negative fluctuations which will tend to cancel each other out. The signal, on the other hand, is nonrandom and can easily be seen on an integrator which adds the signals over a period of time.

I have performed all my calculations with the conditional probability of H_M in the classical order as in equation (26a). As I proved in the

TABLE 4[a]

DNA—RNA INFORMATIONAL PARAMETERS					
Organism or Tissue	*DNA or RNA*	D_1	D_2	R	$RD2$
T2 phage	DNA	.0533	.0057	.0295	.0962
	RNA	.0572	.0118	.0345	.1712
Reovirus	DNA	.0129	.0419	.0274	.7655
	RNA	.0183	.0511	.0347	.7370
Escherichia coli B$_a$	DNA	.0000	.0159	.0080	1.0000
	RNA	.0000	.0184	.0089	1.0000
Micrococcus lysodeikticus	DNA	.1286	.0132	.0709	.0928
	RNA	.1352	.0249	.0801	.1557
Calf thymus	DNA	.0138	.0394	.0266	.7405
	RNA	.0102	.0400	.0251	.7975

[a] Nearest neighbor data for both DNA and RNA of *Reovirus* are from Gomatos *et al.* (1965). All other RNA data are from Weiss and Nakamoto (1961). The DNA values are from Table 3.

self-enclosed section of Chapter 3, it makes no difference theoretically in which direction we write the conditional probability. However, when we deal with the nearest neighbor data, round-off error does accumulate in the third decimal place since this is doubtful in the experimental data. Wherever duplicate data sets for the same organism or tissue exist, I have reported the average values of the information parameters.

The experimental error of the method was determined by duplicate analyses of the same DNA. The standard deviation was then expressed as a percentage of the mean and ranged from 2.33 to 10.0% with an average of 5.8%. When comparing values, I will use the rule-of-thumb that differences greater than 10% of the larger value are unlikely to be experimental errors.

Except for the single case of the satellite DNA of the crab (the light component), these data do not include any values for satellite DNAs but must be assumed to represent primarily the main DNA of the organism. The satellite of the crab is highly redundant DNA. We shall devote Chapter 7 to this type of DNA. I believe that at least some, perhaps all, of the rapidly reassociating DNA of Britten and Kohne (1967) is of this type.

The DNA data which we have today may be the output of a composite source, and future data may make it possible to study separately the individual sources which make up the composite source. A word of caution is in order here. Arrighi *et al.* (1970) present some rather convincing evidence that at least some of the satellite peaks observed in mammalian DNA may be a microbial contaminant, *Mycoplasma*. Therefore we cannot conclude that *any* satellite observed in vertebrate DNA is from a vertebrate source. Much careful experimental work will have to be done in this area.

As shown in Fig. 6, the present data do have reasonably asymptotic behavior. The experimentalist Subak-Sharpe, who has worked extensively with the nearest neighbor data, feels that "the doublet frequencies are reproducible and characteristic of any DNA supplied as template" (Subak-Sharpe, 1969b). Under these conditions one is justified in regarding even a composite source as sufficiently ergodic to justify the calculation of its informational parameters. The ultimate justification is the significance of the results obtained and the role they play in formulating a useful theory which is able to explain a wide variety of empirical phenomena, both past and future. The results we shall obtain in applying the ergodic assumption to the presently existing nearest neighbor data are instrumental in formulating what I hope will be such a theory.

R AND THE *D*-INDICES

Figure 8 is a plot of *R* versus *RD2*. The lower organisms are characterized by a wider variation of *R* versus *RD2* than are the vertebrates. In fact, the vertebrates form a rather localized cluster characterized by both higher *R* and higher *RD2*. Some lower organisms have an *R* value as high as or higher than the vertebrates; but, whenever this occurs, the *RD2* value invariably drops quite low.

FIGURE 8

R VERSUS *RD2*.

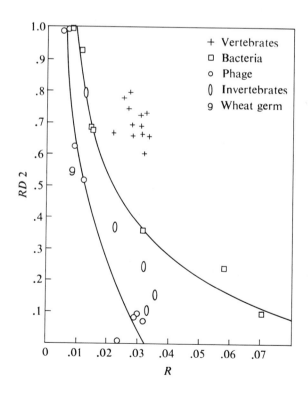

FIGURE 9

R VERSUS *RD*1.

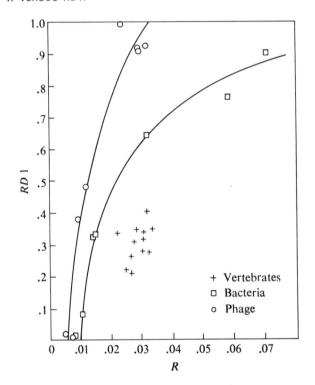

We may describe this situation in another way. Figure 9 is a plot of *R* versus *RD*1. Because $RD1 = 1 - RD2$, this is the graph in Figure 8, rotated. Vertebrates are characterized by higher *R* of the lower *RD*1 type. Table 5 lists the exact ranges of *R* and *RD*1. There are striking differences in the variational ranges of both *R* and *RD*1 values between vertebrates and the lower organisms. *RD*1 runs from 0 to 1 for phage, from 0 to .9 for bacteria, but is restricted to the relatively small range .2 to .4 for vertebrates. *R* runs from .01 to .03 for phage, from .01 to .07 for bacteria, but is confined to the range .02 to .03 for vertebrates. It appears that the vertebrates have set limits on the variation of these parameters.

We may also express this situation in terms of D_1 and D_2. Table 6 lists the ranges of D_1 and D_2; again the vertebrates are characterized

TABLE 5

RANGES OF R AND $RD1$

		R	$RD1$
Phage	Max	.03	.99
	Min	.01	.00
Bacteria	Max	.07	.91
	Min	.01	.00
Vertebrates	Max	.03	$.40^a$
	Min	$.02^b$	.20
Invertebrates	Max	.11	.94
	Min	.01	.21

^{a} .40 = min max. ^{b} .02 = max min.

by a restricted range of D_1 and D_2 values. We will describe this situation in terms of game theory in the next chapter.

D_1 AND D_2

Whenever lower organisms achieve R values in the vertebrate range or above, they do so primarily by increasing D_1, the divergence from equiprobability of the DNA bases. This finding confirms a well-established experimental fact that the base composition of lower organisms, particularly bacteria, has a wide variational range from almost 20 to 80% (C + G) whereas the base composition of vertebrates lies within a relatively restricted range.

Older estimates obtained mostly by chemical methods on the total DNA of an organism or tissue set this range at about 40 to 44%

TABLE 6

RANGES OF D_1 AND D_2

		D_1	D_2
Phage	Max	.059	.015
	Min	.000	.000
Bacteria	Max	.129	.027
	Min	.000	.013
Vertebrates	Max	$.026^a$	.047
	Min	.011	$.029^b$
Invertebrates	Max	.211	.020
	Min	.005	.007

^{a} .026 = min max. ^{b} .029 = max min.

(C + G) (Sueoka, 1964) with a predominance of values at 42%
(C + G). More recent measurements of the base composition of 93
species of mammals by the method of buoyant density in CsCl show
a range of 36.7 to 41.8% (C + G) (Arrighi *et al.*, 1970) for the main
component. These measurements of Arrighi *et al.* (1970) are the first
extensive measurements which take care to separate the main DNA
from the satellite DNA. It appears that the base composition range of
the main peak is somewhat lower than the previous estimates indicated.
There are some comparative values for the same species from both the
older and the newer data. For example, Sueoka (1965) listed 43%
(C + G) for the main DNA of mouse and 33% (C + G) for the
satellite. Arrighi *et al.* (1970) obtained 40.3 to 40.8% (C + G) for
the main DNA of several species of mouse and 23.5 to 30.4% for the
satellite, again somewhat lower.

The base composition range of twelve vertebrate DNAs calculated
(Gatlin, 1966) from the nearest neighbor data (Josse *et al.*, 1961;
Swartz *et al.*, 1962) is 40.5 to 44% (C + G). The values for mouse
thymus and liver are both 41.4% (C + G). It is apparent that the
precise value obtained depends on the particular method used.

We will use the nearest neighbor data as our reference system. These
data indicate that vertebrates have a base composition of about
$42 \pm 2\%$ (C + G) *for the total DNA*. This is a very narrow range
compared to that of the bacteria and lower organisms, which can vary
from almost 20 to 80% (C + G). This narrow base composition range
of vertebrates is a curious fact that has puzzled biologists for many
years. We can now claim that with information theory we have made a
substantial advancement in our understanding of this classical mystery.
The base composition range of vertebrates is an optimum range fixed by
the genetic code. Temple F. Smith (1969) initiated our understanding of
this matter. In Chapter 8 we shall discuss his calculations and our
extension of them.

Our plots of R versus either of the D-indices give us the following
important message:

*Vertebrates have achieved their higher R values by holding D_1 relatively
constant and increasing D_2, whereas the lower organisms which achieve
R values in the vertebrate range or higher do so primarily by increasing
D_1. The mechanism is fundamentally different.*

FIGURE 10

R VERSUS D_2.

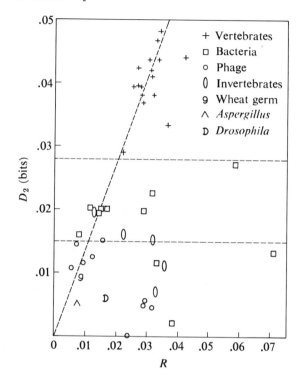

If this is the case, we would expect to find the vertebrate D_2 values higher in general than those for lower organisms. This is what we observe. Figure 10 is a plot of R versus D_2. The vertebrates are characterized by the highest absolute magnitudes of D_2, the divergence from independence of the bases.

We might say that D_2 is an evolutionary index which separates the vertebrates from all other "lower" organisms. In terms of entropy, vertebrate DNA does not necessarily have the lowest values of H_M, but it has the lowest values of H_M relative to H_1, i.e., it has the highest values of D_2—a much more important measure of the ordering of a sequence of symbols than any *single* entropy value.

A word of caution is in order about the biological interpretation of the plot of R versus D_2. It characterizes vertebrates as "higher"

TABLE 7[a]

ADDITIONAL INFORMATIONAL PARAMETERS				
Organism or Tissue	D_1	D_2	R	$RD2$
PHAGE				
Phage α*	.0160	.0158	.0159	.497
BACTERIA				
Rhodospirillum rubrum	.0383	.0195	.0289	.338
Serratia marcescens*	.0133	.0204	.0169	.606
Proteus vulgaris	.0530	.0137	.0333	.205
Bacillus megaterium	.0740	.0043	.0391	.055
PLANT				
Aspergillus nidulans	.0100	.0053	.0077	.346
INSECT				
Drosophila melanogaster	.0281	.0060	.0170	.175
VERTEBRATE				
Rana catesbeiana*	.0152	.0396	.0274	.723
Mouse satellite	.0624	.0454	.0539	.421
Mouse main band*	.0189	.0482	.0336	.718
BHK 21/C13	.0389	.0336	.0363	.463
BHK 21	.0391	.0439	.0415	.529

[a] BHK 21/C13 is from Morrison *et al.* (1967). BHK 21 is from Subak-Sharpe *et al.* (1966). All other animal and plant data are from McGeoch (1970). Virus data are as follows:
  1–8: McGeoch (1970).
  9–14: Subak-Sharpe *et al.* (1966).

organisms on the basis of an information theoretic measure. There is no particular reason to expect this measure to correlate *in detail* with a classical biological taxonomic measure. For example, the phage and bacteria seem to aggregate disjointly, whereas the invertebrates overlap both categories. The invertebrate phyla represented are Protozoa, Echinodermata, and Arthropoda. In terms of classical taxonomy, invertebrates should be "higher" organisms than either phage or bacteria. However, from an information theoretic viewpoint, we may regard them all as "lower" organisms because of the lower degree of organization of their DNA base sequences. The data are limited. Important groups such as insects are not even represented. It may be

TABLE 7 (Continued)

Organism or Tissue	D_1	D_2	R	RD2
VIRUSES				
1 Ad-2	.0031	.0102	.0066	.767
2 Ad-4	.0053	.0122	.0087	.697
3 Ad-7	—0—	.0088	.0035	1.000
4 Ad-11	.0025	.0106	.0065	.809
5 Ad-12	.0125	.0196	.0161	.610
6 Ad-18	.0094	.0138	.0116	.594
7 Ad-21	.0013	.0156	.0085	.923
8 Ad-27	.0066	.0206	.0136	.757
9 Herpes	.0555	—0—	.0250	—0—
10 Pseudorabies	.1451	.0022	.0736	.015
11 Polyoma	.0080	.0328	.0204	.803
12 Equine rhinopneumonitis	.0037	.0011	.0024	.265
13 Vaccinia	.0783	.0031	.0407	.042
14 Shope papilloma	.0090	.0242	.0166	.729
15 Human papilloma	.0220	.0167	.0194	.431
16 Simian virus 40	.0361	.0564	.0462	.670
17 Encephalomyocarditis	.0079	.0478	.0278	.859
18 H-1	.0085	.0561	.0323	.869
19 Kilham rat virus	.0134	.0380	.0257	.739
20 Minute virus of mice	.0306	.0504	.0405	.621
21 Newcastle disease virus	.0053	.0218	.0136	.804

15–16: Morrison et al. (1967).
   17: Hay and Subak-Sharpe (1968).
18–20: McGeoch et al. (1970).
   21: Scholtissek and Rott (1969).
* This falls on the RD2 vector.

that they belong in the vertebrate domain under the D_2 measure. We will attempt to show in the section on Shannon's second theorem more precisely what we mean by "higher" and "lower" organisms.

On the other hand, there should be a broad correspondence between our classical notions of complexity of the organism and the D_2 measure.

After the entire preceding portion of this chapter was written and in press, a substantial new set of data was brought to my attention by Robert A. Elton of Subak-Sharpe's group. I could have incorporated this data into Table 3, but I thought it would be more interesting to the reader to present it as data testing a theory already formulated. Table 7 contains the additional data and the points for animals and plants have

FIGURE 11

R VERSUS D_2 FOR MAMMALIAN VIRUSES.

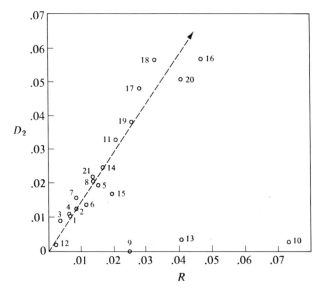

been added to Fig. 10. In Fig. 10, phage α overlaps slightly into the previously defined bacterial domain and two of the bacteria, *Proteus vulgaris* and *Bacillus megaterium*, fall into the phage domain. The single insect, *Drosophila melanogaster*, is low in the phage domain as is the additional plant, *Aspergillus nidulans*. The two BHK21 data sets represent hamster cells grown in tissue culture. Hence we would not necessarily expect them to represent normal vertebrate tissue. However even these points are well within the vertebrate range. They do fall off the *RD2* vector, which we shall discuss presently.

The mouse satellite is an *R*-redundant sequence with a short repeat length. We shall discuss such DNA in Chapter 7.

Table 7 contains 21 mammalian viruses. A separate plot of D_2 versus *R* is shown in Fig. 11. These viruses represent a rather heterogeneous group. Some are DNA, some RNA; some are double-stranded, some single stranded; some are large, some small; some are oncogenic, some non-oncogenic. There is no evidence to indicate that a similarity to mammalian *R* and *D* indices is either a necessary or sufficient condition for oncogenicity. However, there does appear to be a general correlation between the two. The possibility exists that a necessary (but not sufficient) condition for oncogenicity is that the virus lie on the *RD2* vector. None of the viruses which fall off the vector are oncogenic.

Tables 3 and 7 contain 46 organisms in all. In not a single case do we find a lower organism with a D_2 value in the vertebrate range, or vice versa. Throughout the remainder of the book I will use only the data set of Table 3, because none of my conclusions have been altered by the new data.

OTHER EVOLUTIONARY INDICES

A statistician might be interested in whether or not the result that vertebrates have the highest values of D_2 could be duplicated by classical statistical procedures. After all, D_2 is a measure of the "deviation from random" of the base sequence in DNA. Usually, when a statistician speaks of a "random" sequence, he means one where there has been no divergence from independence of the symbols as separate and distinct from the divergence from equiprobability.

Figure 12 is a plot of R versus σ, where

$$\sigma = \left\{ \sum_i \sum_j \frac{(p_i p_{ij} - p_i p_j)^2}{16} \right\}^{1/2} \tag{61}$$

FIGURE 12

R VERSUS σ.

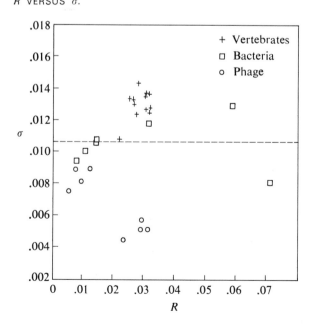

σ is the standard deviation of the difference between the real doublet frequencies and the "random" doublet frequencies, taking into consideration only a first-order Markov dependence. This should be a classical counterpart of our D_2 information theoretic measure. However, σ cannot begin to duplicate the results of the D_2 index. There is significant overlap of the bacterial and vertebrate domains.

It is possible to define arbitrarily other classical evolutionary indices by use of the standard root mean square form with a slightly different "base." Figure 13 is a plot of R versus e, where e is a classical evolutionary index defined by Temple F. Smith (1969):

$$e = \left\{ \sum_i \sum_j \frac{(p_{ij} - p_j)^2}{16} \right\}^{1/2} \tag{62}$$

Here the "base" of the index is the transition matrix element p_{ij}

FIGURE 13

R VERSUS e

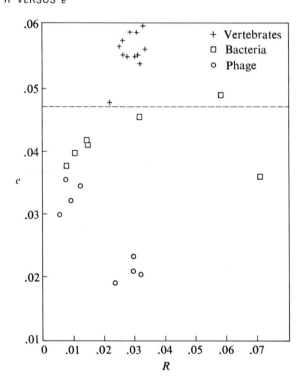

minus the base composition value p_j. The results are much better. There is separation of the vertebrate and lower organism domains with only a very slight overlap at the boundary.

Subak-Sharpe *et al.* (1966) make the following intuitive manipulation with the nearest neighbor data. First the real doublet frequency is divided by the "random" doublet frequency. Since this ratio is near unity, it is divided by 16, the total number of doublets, to obtain a "normalized" doublet frequency which Subak-Sharpe regards as the doublet frequency that the organism would have if its DNA bases were equiprobable.

Subak-Sharpe (1969a) speaks of the pattern of normalized frequencies as "general designs" of nucleic acids. He states: "Strictly speaking the general design of a nucleic acid is the pattern of deviations from random expectation for the sixteen doublets." Usually, when a biologist uses the phrase "deviation from the random" in describing a sequence of symbols such as DNA or RNA, he, like the statistician, means the divergence from independence of the symbols as separate and distinct from the divergence from equiprobability. This is precisely what we have been calculating in D_2 and D_1, respectively. But it should be understood that D_1 is a "deviation from the random" also because the most "random" state is the maximum entropy state and this is characterized by events which are both independent and equiprobable. Usually, however, when dealing with sequences, we use the word random to describe independent events, and the phrase "deviation from the random" to refer only to D_2.

Subak-Sharpe's procedure is an intuitive attempt to conserve some invariant property of the doublet frequency matrix. Since the attempt obviously is intended to remove the contribution of the base composition D_1, it is clear that D_2, the divergence from independence, is the desired invariant.

Of course, D_2 can be calculated for a given organism; we may then ask the question, "What are the doublet frequencies with this value of D_2 but with $D_1 = 0$?" However, there is no algorithm or decision procedure which can answer this question. The problem may be formulated as follows. Since $D_2 = H_1 - H_M$,

$$D_2 = \sum_j \sum p_i p_{ij} \log \frac{p_{ij}}{p_i} \tag{63}$$

Given the desired value of D_2 and given the p_i, equation (63) contains 16 unknown values of the p_{ij}. The Watson-Crick equivalence relations along with the known base composition reduce the number of independent transition matrix elements to 6. This leaves us with one equation in 6 unknowns.

Therefore there is no algorithm, no finite number of manipulative steps, whereby one can "normalize" or transform a doublet frequency matrix to the values it would have at $D_1 = 0$ which conserves D_2. One can easily verify that Subak-Sharpe's algorithm does not preserve D_2. Therefore one should not regard Subak-Sharpe's normalized frequencies as the doublet frequencies which the organism would have at 50% (C + G). This interpretation is mathematically unsound. If one wished to estimate these frequencies, he should utilize optimization techniques (Bremmerman and Lam, 1970) to minimize the difference between the value of D_2 from estimated frequencies and the desired value of D_2. This can be done.

However, even though this algorithm is not being interpreted correctly, it has other interesting properties. The "normalized" frequencies are never used except as an intermediate computational point. The algorithm next subtracts the "normalized" frequencies from the "random" frequencies at 50% (C + G), namely, 1/16. The entire Subak-Sharpe algorithm may be summarized in the following index.

$$SS = \frac{1}{16} - \frac{p_i p_{ij}}{16 p_i p_j} \tag{64}$$

or

$$SS = \frac{1}{16}\left(1 - \frac{p_{ij}}{p_j}\right) \tag{65}$$

Let us define the following evolutionary index:

$$SSe = \left\{ \sum_i \sum_j \frac{1}{16}\left(1 - \frac{p_{ij}}{p_j}\right)^2 \right\}^{1/2} \tag{66}$$

In this index the quantity $(1 - p_{ij}/p_j)$ replaces the quantity $(p_{ij} - p_j)$ in Smith's e index. Figure 14 is a plot of R versus SSe. The pattern of separation of phage, bacteria, and vertebrates is identical with the plot of R versus e.

FIGURE 14

R VERSUS *SSe*.

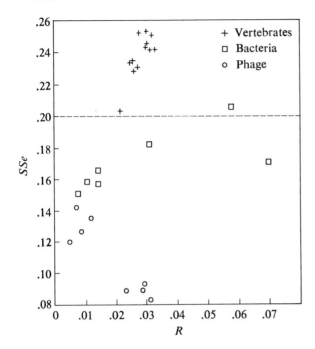

Both of these evolutionary indices are reasonably good because both utilize only certain properties of the transition matrix of the Markov chain, rather than the doublet frequency matrix. The *e* index is based on the difference of the real and the random transition matrix elements, whereas the *SSe* index utilizes one minus the ratio of the real to the random transition matrix elements. Both of these quantities must be functions relatively free of the influence of D_1 or the base composition because the evolutionary indices defined utilizing these quantities both come close to mimicking the evolutionary index D_2. However, they are not mathematically equivalent because nowhere in the definition of either *e* or *SSe* does the concept of entropy enter in along with its inevitable logarithmic functional form.

I feel that D_2 is superior as an evolutionary index. The reasons for this bear an interesting analogy to a chapter in the history of physics. Under Newton's laws of motion, the path of a planet about the sun is required to be an ellipse. The perihelion is that point in the path which is nearest the sun. It was known even under Newtonian mechanics that the perihelion of the path of the planet Mercury about the sun would shift a little with each revolution as a result of the "perturbations" caused by the presence of other planets. This shift in the perihelion of Mercury could be calculated as 531 seconds of arc per century, whereas the observed shift was actually 574 seconds per century, leaving a discrepancy of 43 seconds per century unaccounted for by Newtonian mechanics. Einstein's general theory of relativity accurately accounted for this discrepancy of 43 seconds of arc per century, yet no physicist would argue that general relativity is superior to Newtonian mechanics because of this slight improvement in the quantitative description of a natural phenomenon. The value of relativity was in the vast domain of conceptual richness and beauty which it opened up. It expanded the horizons of physics on a scale which no one could have predicted from the magnitude of the shift in the perihelion of Mercury.

The information theory evolutionary index D_2 is slightly better quantitatively than any of the classically defined indices. There is no overlap between the vertebrates and bacteria. Perhaps this result is important. The history of physics has many instances of the importance of small quantitative differences. On the other hand, perhaps the differences are best attributed to experimental error. I will not argue the point. This is not the primary reason why D_2 is vastly superior as an evolutionary index. The reasons are conceptual. D_2 is an entropy function. It extends the entropy concept and endows it with structure. It describes the DNA base sequences of vertebrates as more highly ordered than those of lower organisms. We are left not with merely an isolated, arbitrary result, as we would have been with the classically defined evolutionary indices, but with an explanation of our result and a workable theory which allows us to explore a vast and new conceptual area.

THE RD2 VECTOR

In 1968 I published the plot of R versus D_2 (Fig. 10) for the data set of Table 3. The data as a whole is not linear. However, after reading Reichert and Wong's (1971) paper, which shows a distinct linearity for

R versus D_2 on the protein space of cytochrome c under their information measures, I examined Fig. 10 again. The vertebrate cluster does
have a definite linear bias. If we draw a line through this cluster, it
intersects the origin. Thus in equation (53), $\omega = 0$. $\omega = 0$ also for the
plot of Reichert and Wong. Some of the lower organisms lie on this line
(drawn in Fig. 10) and some fall off it. The linearity of the plot of R
versus D_2 is much more apparent for the mammalian viruses. This is
shown in Fig. 11. Here, too, $\omega = 0$. As we noted previously, the coordinates (R, D_2) define a vector, the $RD2$ vector.

The existence of the $RD2$ vector can be predicted from a simple
random generator experiment. I have generated random sequences
from an alphabet of 4 letters from length 4 to 100. Each sequence of a
given length was generated 100 times with the Berkeley CDC 6400
system generator, and the average value of the information parameters
was calculated. Figure 15 shows a plot of D_1 and D_2 versus the length of

FIGURE 15

D_1 AND D_2 VERSUS LENGTH OF SEQUENCE
FOR DNA RANDOM GENERATOR.

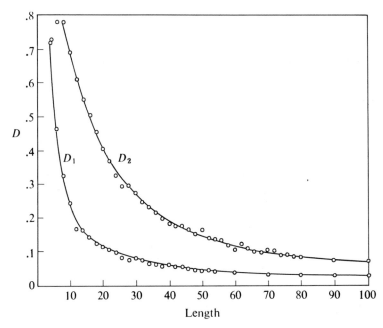

FIGURE 16

R VERSUS D_2 FOR DNA RANDOM GENERATOR.

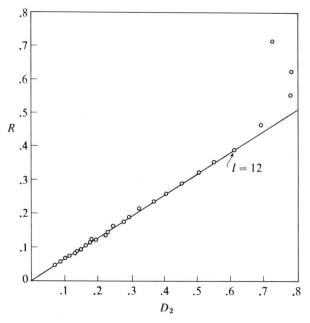

the sequence. D_1 drops off more rapidly than D_2 and approaches a stable value asymptotically. This is because the size of S_1 is small relative to S_2 and the distribution of the p_i on S_1 approaches equiprobability the most rapidly; hence the D_1 value stabilizes first.

Figure 16 is a plot of D_2 versus R. At relatively short lengths (about 12) the plot becomes linear. The theoretically predicted slope for $D_1 = $ constant is $1/\log a$ or $1/2$. Its value in Fig. 16 is .645. This occurs because D_1 is still declining slowly in accordance with equation (54). Thus the random generator predicts that $\alpha = .645$, $\omega = 0$ for DNA. For both Figs. 10 and 11, $\alpha = .7$, $\omega = 0$.

Note that before length 12, the R and D_2 functions do not follow any regular functional form. I have included these points on Figs. 15 and 16 but have not included them in the functional curve or line. Thomas (1966) has estimated that the "minimum recognition length" for stable complementary nucleic acid strand union is greater than or equal to 12. This is explainable in the language of our theory. We shall note in the next section that D_1 must "stabilize" before language formulation is possible. We can be more explicit. After R versus D_2 enters its linear phase, D_1 versus D_2 also becomes linear. For nucleic acids the change occurs at about length 12. This is truly a fundamental "minimum recognition length" for pattern recognition in nucleic acids.

The random generator predicts the $RD2$ vector but does not predict its *direction*. In fact, it predicts that the redundancy of a sequence of symbols should decline with increasing length. However we know that as higher organisms have evolved, there has been a steady increase in the amount of DNA per cell (cf. Britten and Davidson, 1969). More specifically, higher organisms have maximized the minimum amount of DNA per cell. We shall discuss this in Chapter 7. If we regard the DNA of a cell as a single molecule specifying the organism, there has been an increase in the length of the DNA of higher organisms.

Thus the evolution of life goes in exactly the *opposite* direction to that predicted by the random generator. In this sense, the $RD2$ vector is "truly time's arrow" as Reichert and Wong suggest. We will discuss this in more detail in Chapter 9.

In my previous work and in discussion up to this point in the book, I have usually included the word *relatively* in the phrase "increasing D_2 at *relatively* constant D_1." I detailed the base composition range of vertebrates and noted that it is not an *exact* constant. As equations (53) and (54) show, it is the *slope* of the $RD2$ vector which tells us how nearly constant D_1 is on a given information space. When $\alpha \neq 1/\log a$, then as D_2 increases, D_1 also increases in accordance with the linear relation, equation (54). The significant point, however, about the evolution of vertebrates from lower organisms is that this increase of D_1 has been confined within very narrow limits.

This must be the case. Otherwise the D_1 value will fall outside the optimum range set by the genetic code. We shall derive this range in Chapter 8. The organisms ancestral to the vertebrates must have confined D_1 to this range. All such organisms would fall on or near the $RD2$ vector. Those organisms which fall far off the vector have allowed D_1 to vary too much. We can show that this is poor strategy from the basic principles of language formulation.

EVOLUTION OF THE GENETIC LANGUAGE

Let us assume that the first DNA molecules assembled in the primordial soup were random sequences, i.e., D_2 was zero and also D_1.

One of the primary requisites of a living system is that it reproduce itself accurately. If this reproduction is highly inaccurate, the system has not survived. Therefore, any device for increasing the fidelity of information processing would be extremely valuable in the emergence of living systems.

Redundancy, or information density, is a measure of all the constraints placed upon a sequence which make possible error detection and correction. Therefore redundancy is in this sense a measure of the fidelity of a message. Some lower organisms first attempted to increase the fidelity of the genetic message by increasing R primarily by increasing D_1, the divergence from equiprobability of the symbols. This is a very unsuccessful and naive technique because, as D_1 increases, the potential message variety declines. We may show this quantitatively as follows.

As the base composition deviates from the central mean of 50% $(C + G)$, D_1 increases until, in the limit, D_1 takes on its maximum value, log a. We have then a monotone, a sequence of only one letter, where the probability of one letter is 1 and of all others 0. Clearly there is no message variety at all in such a sequence. Thus the potential message variety declines steadily from a maximum at 50% $(C + G)$ to zero at the monotone.

It is not difficult to show this. Let us define the potential message variety, *PMV*, of a sequence as the maximum possible number of different words of fixed length, l, that can be formed per unit message length, la^l, for a given value of D_1 where a is, as before, the number of letters in the alphabet, a^l is the maximum number of different possible words of length l that can be formed assuming no sequencing restrictions, and la^l is the minimum length of the sequence just necessary to list each word once.

An example shows the decline of *PMV* with increasing D_1 for the simplest possible case; Let $a = 2$, $l = 2$:

Words		D_1 (bits)	*PMV* (words)
AT TA TT AA		0	4
TT TA TT AA	(replace A by T)	0.05	3
TT TA TT AT		0.19	3
TT TT TT AT		0.46	2
TT TT TT TT		1.00	1

Similarly we can show for higher values of a and l the decline of PMV with increasing D_1. D_1 is a continuous function, whereas PMV is a step function.

Please note that throughout the discussion above I have been careful to prefix the word "potential" before each usage of the term "message variety." This dichotomy of potential versus the utilization of that potential is applicable to both conjugate elements of the information concept. Strictly speaking, redundancy measures only a capacity to combat error, not the actual utilization of that capacity, and equi-probability of the symbols ($D_1 = 0$) measures only a capacity for maximal message variety. This does not necessarily require that the actual or real message variety be maximal. We may define the real message variety RMV as the number of different words of length l actually encountered when a unit sequence of length la^l is read in units of l from either terminal end.

In the example, when $D_1 = 0$, it is possible to rearrange the letters such that $RMV = 1$ although $PMV = 4$. For the sequence

ATCGATCGATCG . . .

$D_1 = 0$. If $l = 3$, $PMV = 64$, whereas $RMV = 4$.

Although PMV is a function of D_1 only, RMV is a function of both D_1 and D_2. As D_2 increases, RMV ultimately declines and in the limit is a highly ordered sequence such as the example we have just given. This is why we always place the phrase "within limits" or "under certain conditions" in the statement of Shannon's second theorem, which we shall encounter presently.

Therefore, if an organism attempts to achieve a higher R value simply by increasing D_1, it will pay too high a price in loss of potential message variety. This is what happens to the lower organisms which fall off the $RD2$ vector.

A much more sophisticated technique for increasing the accuracy of the genetic message without paying such a high price for it was first achieved by vertebrates and their ancestors.

First they restricted the range of D_1. This is a fundamental pre-requisite to the formulation of any language, particularly more complex languages. We observe it in human languages. The particular distribution of the single letter frequencies in human language is so stable and characteristic of a given language that this is a fundamental tool used by cryptographers in decoding messages.

When a cryptographer is faced with an unknown message, he first begins to count the single letter frequencies. If the message is in English, the letter e will always be the most frequently occurring, provided the text is of sufficient length. The distribution of the p_i on S_1 is stable and characteristic of a given language. The vertebrates and their ancestors were the first living organisms to achieve the stabilization of D_1, thus laying the foundation for the formulation of a genetic language. Then they increased D_2 while allowing D_1 to increase from zero only up to the optimal boundaries fixed by the genetic code. Hence, they increased the reliability of the genetic message without a great loss of potential message variety. They achieved a reduction in error probability without paying too great a price for it, and an information theorist would recognize this as the utilization of Shannon's second theorem, the coding theorem of information theory.

SHANNON'S SECOND THEOREM

Without question, the climax of all information theory is the remarkable second theorem of Shannon. While this is usually expressed in highly technical, mathematical terms, its underlying concept is not difficult to grasp; but first we must define some simple terms from the jargon of the communications engineer. See the engineer's diagram in Fig. 17.

A source or transmitter is any apparatus that emits a sequence of symbols. As we have noted, when this sequence of symbols is ordered

FIGURE 17

THE ENGINEER'S DIAGRAM.

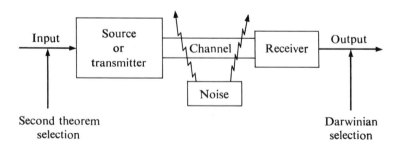

according to a set of constraints, the set of constraints constitutes a language and the sequence of symbols itself is called the message. The encoding of the message in a particular language occurs at the source. A channel is simply any medium over which the message is transmitted. For example, it may be some simple physical device like a television cable or it may be a rather complicated system such as a computer. Conceptually, it is anything one regards as intermediate between the transmitter and receiver and, hence, may sometimes be somewhat a matter of definition. In the living system, we will regard the base sequence of DNA as the encoded message from the source and the amino acid sequence in protein as the message which is finally received at the output. The channel consists of the entire mechanics of protein synthesis involving transcription of the base sequence of DNA into messenger RNA and the translation of the messenger into protein on the ribosome.

The channel has a certain capacity, an upper limit to the rate at which it can transmit a sequence of symbols without incurring gross error due to overloading the channel. Again there is a precise mathematical definition of channel capacity, but the intuitive concept is obvious and adequate for our purpose.

In the channel, noise may interfere with the transmission process. Specifically, any malfunction of the transmission mechanics which causes error in the message received may be regarded as noise. It is impossible to eliminate completely the noise in any real system, especially one as complicated as the living system, and it is at this point that the second theorem of Shannon intervenes. The essential concept is this: we cannot eliminate noise in the channel, but we can under certain conditions transmit a message without error and without loss of message rate in spite of this noise *if the message has been properly encoded at the source.* The code is the crux of the matter. The more efficient the code becomes, the closer it approaches this error-free limit. I will use the word efficiency in this sense: the greater the efficiency, the lower the error.

Shannon's theorem proves that, if the source and channel have certain properties, then for long messages a code exists which can reduce the error to an arbitrarily small value even in a noisy channel. The most obvious way to ensure that a message will get through a noisy channel is simply to repeat it many times. However, repetitive redundancy always purchases reliable transmission at the cost of reduced transmission rate. The incredible thing about the second theorem is that it promises error-free transmission in noisy channels *without reduction in transmission rate*; and the upper limit to this error-free transmission rate is the channel capacity. The second theorem tells us that all this is possible if only we can find precisely the right code. It does not tell us *how* to find this code. The theorem, like so many theorems of higher mathematics, simply tells us that such an entity *exists*. The search for general methods of constructing such codes has long been the Holy Grail of information theory (Berlekamp, 1968), and it is just possible that the living organism in its quest for evolutionary advancement has the same goal.

Let us examine some of the conditions under which this almost incredible promise is made. We noted that any channel has a certain capacity or upper limit to the rate at which it can transmit a sequence of symbols. If the rate at which the source emits its symbols exceeds this capacity, it is clear that chaos will result. Hence a fundamental condition under which the Shannon theorem is valid is that the rate of emission from the source, which is measured by the source entropy, must not exceed the channel capacity. I assume that life could not have evolved if this condition had not been met, i.e., life could not have evolved on an overloaded channel. A second fundamental condition is that the source be ergodic. We discussed this concept in the section on ergodic Markov sources in Chaper 3.

This entire statement of the second theorem principle is still imprisoned in the jargon of the communications engineer. I make the following generalization. I regard the uncertainty principle as the primitive root of the entire information concept. We described at length the two opposing elements, variety versus reliability. Variable pairs which represent mutually antagonistic elements are called canonically conjugate pairs in quantum theory. The two opposing elements of the information concept are in a primitive sense analogous to the canonically conjugate variable pairs of Heisenberg's uncertainty principle. Weaver (1949) puts it succinctly when he states that the opposing elements of conjugate variable pairs are "subject to some mysterious joint restriction that condemns us to the sacrifice of one as we insist on having more of the other." The information conjugate variables are certainly subject to such a restriction because one of them, variety, is ultimately amplified by increasing the entropy while the other, reliability, is ultimately amplified by decreasing the entropy.

However, with the introduction of the D-indices and Shannon's second theorem this situation takes on a whole new dimension. First of all, the terms "increasing entropy" and "decreasing entropy" are no longer completely definitive as they were in classical thermo-dynamics. We must now ask *how* the entropy was increased or decreased. What was the relative variation of D_1 and D_2? What are the final D-indices? Under the second theorem principle it is possible to "have your cake and eat it too" within limits. It is possible to increase the reliability of a message without loss of message variety *if* the entropy is reduced in a particular way, namely, by increasing D_2 at relatively constant D_1.

There are limits to this process. If the entropy continues to decline by any mechanism, the actual message variety will ultimately decline. If D_1 is maximal, we would have a monotone as in our previous example and, even if $D_1 = 0$, as D_2 approaches its maximum value, log a, the actual message variety will decline. Capacity is of no value if it cannot be utilized. The uncertainty principle restricts us to the ultimate sacrifice of actual message variety as we insist on having more actual error reduction. This restriction becomes more and more confining as the system approaches extreme values of its entropy variables. *However, the second theorem principle declares a moratorium on the uncertainty restriction within limited ranges of the entropy variables, provided that the variables change in the proper way. This is simply a generalized formulation of the engineer's principle that we can, under certain restrictions, reduce the error in transmission of a message through a noisy channel without loss of transmission rate, provided that the message has been properly encoded at its source.*

Now let us state the second theorem in the language of the biologist. It is possible, within limits, to increase the fidelity of the genetic message without loss of potential message variety, provided that the entropy variables change in the proper way, namely, by increasing D_2 at rela-tively constant D_1. This is what the vertebrates have done. This is why we are "higher" organisms.

LANGUAGE IN GENERAL

Let us look back and evaluate what we have accomplished up to this point. In review, the theory upon which the definition of D_1 and D_2 is based is perfectly general and could be applied to language in general. Then we observed in the genetic language the increase of D_2 at relatively constant D_1 as a fundamental mechanism for increasing the fidelity of the genetic message. Now I ask the questions: "Is this mechanism a general mechanism for increasing the fidelity of any message? Is it used anywhere in human language?" It is.

The human mind is, after all, an information processing channel, the most complex in the universe as far as we know, and like any channel possesses a certain capacity, an upper limit to the rate at which it can receive and process information. If information is transmitted at a rate which overloads this capacity, the result is *not* that an amount of information up to the channel capacity is received and processed and the rest "spills over." The result of overloading the channel is utter confusion and chaos. Any good teacher knows this, and very carefully and with deliberation lays a firm foundation of fundamentals before increasing the rate of transmission of information to the student. It is extremely important in the initial stage of this process that error be held to an absolute minimum. Therefore any device for increasing the fidelity of a message is extremely useful.

One of the most important learning processes which the human mind undergoes is that when a little child learns to read the written language. He has spoken it for several years before he learns to read it, and this is a major advancement. It is obvious that any safeguards against error in the early critical stages of this learning process would be invaluable.

I shall now show that the writers of children's textbooks intuitively utilize the basic device of increasing D_2 at relatively constant D_1 to increase the fidelity of the message. I selected a series of well-known children's readers beginning with the primer and continuing through the sixth grade. The series selected is the Ginn Basic Reader series (Ginn and Company, Boston). I calculated the redundancy of each book by taking texts of increasing length until the R value stabilized, taking into consideration only the first-order Markov effect. Figure 18 is a plot of R versus the grade of the reader. R is quite high in the primer and follows a very smoothly declining curve as the grade of the reader increases.

FIGURE 18

R VERSUS GRADE OF READER.

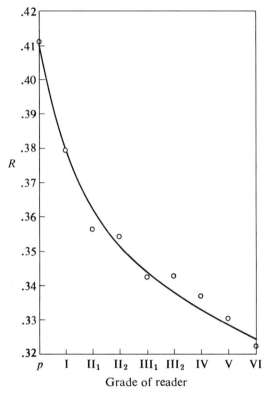

Grade of reader

In Fig. 19 I have taken the R value apart into D_1 and D_2. When we do this, it is very apparent that the high R value in the early readers has been achieved by increasing D_2 at relatively constant D_1, just as the vertebrates did. Therefore this appears to be a fundamental mechanism, a general mechanism, for increasing the fidelity of a message.

This result is very reassuring. It shows the strategy of increasing D_2 at relatively constant D_1 in two languages, the genetic language and the human language. In fact, the word "strategy" is perhaps the best way to describe it. We shall devote the entire chapter which follows to this view.

INFORMATION SPACES

The $RD2$ vector describes "the path by which the system is required to climb" (Eigen, 1971) if it is to optimize the conflicting

FIGURE 19

D_1 AND D_2 VERSUS GRADE OF READER.

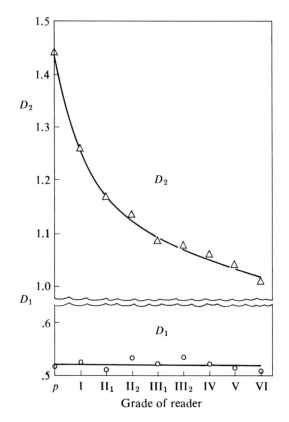

elements of variety versus reliability. It is an optimum path. I suggest that the organisms ancestral to vertebrates climbed this path. This does not necessarily mean that the organisms now living which we observe on this line were *the* ancestral organisms. Restriction to the vector is a necessary but not sufficient condition. However, the properties of the vector should tell us something about the nature of the type of organisms which gave rise to vertebrates.

The $RD2$ vector is a fundamental invariant, a tensor of rank one. We shall confine the following discussion to the particular case where $\omega = 0$ since this is the only case observed in nature to date. Therefore we shall be particularly interested in the coordinate transformations involving a simple rotation of the coordinate axes about the origin.

Since the $RD2$ vector is invariant under this coordinate transformation, the R and D_2 axes can always be rotated until the slope of the vector equals $1/\log a$. Thus an information space always exists where D_1 is a constant and we may regard advancement along the $RD2$ vector as a process of increasing D_2 at constant D_1. We have already encountered an example of this in human language and we shall observe it again in Chapter 8 in our calculations of the capacity of the DNA-to-protein channel. In both cases the plot of R versus D_2 has a slope of $1/\log a$.

We will call the information space of Fig. 10 "DNA space" because it is derived from the experimental nearest neighbor data. In DNA space, the $RD2$ vector has the following properties. For DNA, $\log_2 a = 2$ bits. Using this metric, from (54) and (59) we have

$$RD2 = \frac{1}{2\alpha} \tag{67}$$

From Figs. 10 and 11, $\alpha = .7$. Hence $D_1 = .4D_2$ and $R = .7D_2$. However, we have the symmetry property

$$\alpha \cong \frac{1}{2\alpha} \tag{68}$$

or $\alpha = \sqrt{.5} \cong .7$. Hence

$$RD2 = \frac{1}{2\alpha} \cong \alpha \tag{69}$$

Thus in DNA space, the slope of the $RD2$ vector is approximately equal to the $RD2$ index. All the organisms which fall off the vector have $RD2$ values significantly different from α. In fact, they are all less than .5. In Tables 3 and 7 I have indicated all the lower organisms that lie on or near the vector. (The vertebrates all lie on it except for the two BHK21 points.)

If we make a coordinate transformation to the space where $\omega' = 0$, $\alpha = 1$, we have under the $\log_2 a = 2$ metric, $RD2 = .5$ and

$$D_1 = D_2 = R \tag{70}$$

This is a perfectly symmetric space. If we plot $RD2$ versus $\%\,(C + G)$ as shown in Fig. 20, most of the experimental points for lower organisms

FIGURE 20

*RD*2 VERSUS %(C + G).

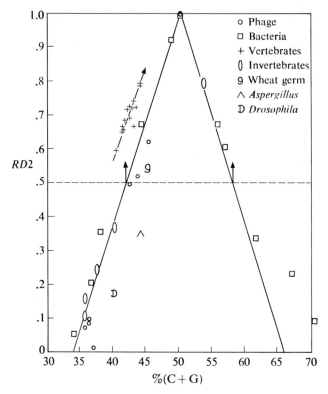

lie on or near two lines symmetric about the central random plane of
50% (C + G). The vertebrate values lie on or near a parallel vector in
the base composition range of 42 ± 2% (C + G).

It is interesting that the line on the vertebrate side of the random
plane intersects the line $RD2 = .5$ at exactly 42% (C + G), the optimum
base composition of the code. At this point under the base 2 metric,
$D_1 = D_2 = R$. This is a "balance point" expressing the symmetry of the
genetic code on this space and under this metric. It was somewhat
disappointing that the genetic code displayed no particular symmetry as
satisfying as that of DNA structure. However, a very satisfying symmetry exists on this information space.

Figure 20 and equation (70) constitute the first mathematical evidence

I have seen which suggests (but does not prove) that the present genetic code is an optimum code rather than an arbitrary code. The reasons for this optimization must be sought in mathematical terms rather than in any physical interaction between amino acids and nucleic acids.

It is irresistibly tempting to speculate at this point. *Let it be clearly understood that the following paragraph is not a scientific discussion* but rather the type of speculative fantasy which all scientists enjoy from time to time. We never print them in scientific journals, but they make life interesting.

The organisms which lie on or near the $RD2$ vector are the most likely candidates for organisms ancestral to the vertebrates. From Fig. 20 we see that symmetry is also an ancestral prerequisite. The two lines are symmetric with respect to reflection in the random 50% $(C + G)$ plane. The vertebrates have arisen only on one side of this plane. There are organisms which lie on the vector, but do not have the proper symmetry, i.e., they lie on the wrong side of the random plane. Hence these are not ancestral to the vertebrates. One bacterium, *Bacillus subtilis*, and 3 phage, phage α, ϕX-174, and T1, fall on the $RD2$ vector and have in addition the proper symmetry. They also have the optimum base composition. From such organisms the vertebrates have arisen. As shown in Fig. 20 by the two small vertical arrows, there are two balance points possible, one at $\approx 42\%$ $(C + G)$ and the other at $\approx 58\%$ $(C + G)$. Vertebrates have arisen at only one of them. What would higher organisms be like which originate at the other balance point, the point which has the mirror image symmetry of vertebrates? We do not find such beings on this planet; but then neither do we find anti-matter on this planet. Yet physicists believe that anti-matter exists somewhere in the universe. Physicists have their matter and anti-matter. Biologists can now talk about life and anti-life. We can now define vertebrates and anti-vertebrates.

5

There lies a green field between the scholar and the poet; should the scholar cross it he becomes a wise man; should the poet cross it, he becomes a prophet.—Kahlil Gibran, Sand and Foam

THE HIERARCHY AND THE GAME

ARTHUR KOESTLER

In a broad sense we may regard a scientist as being on one side of Kahlil Gibran's green field and a creative writer as on the other. One of the writers of our time who has crossed this field of which Gibran speaks is Arthur Koestler. In his book, *The Ghost in the Machine*, he presents what I feel is a penetrating analysis of the hierarchy concept. This concept is vital to Polanyi's philosophy, and in this chapter we shall attempt to blend the hierarchy concept with the concepts of game theory.

In any complex system designed according to the pattern of increasing levels of complexity, ordered levels of control layered one on top of the other, structure upon structure upon structure, we find that the words "whole" and "part" are completely relative. For example, the vertebrate heart functions as a unit. It is composed of parts, the right and left auricles, the right and left ventricles, which function harmoniously as a whole. Yet the heart is only part of the whole body. Man is a complete biological whole, but in relation to society in general he is a very small part of the complex social system in which he functions. Koestler (1967) has coined the word "holon" for these functional units which are parts relative to the higher levels of organization and wholes relative to the lower levels.

Holons are torn between two opposing forces which Koestler terms the self-assertive tendency and the integrative tendency. These two faces of Janus are none other than the two opposing elements of the information concept, freedom versus contraint, which have played a central role in our thinking. The self-assertive tendency is the drive of the holon to express its wholeness with complete freedom from constraint of any kind. The integrative tendency is the drive to become a part of a larger whole, which entails acceptance of restriction of personal freedom. A hierarchy cannot function unless its parts obey the higher laws. For example, in higher organisms all the vital functions of metabolism, circulation, respiration, etc., function automatically. We could not survive if each of these functions required detailed conscious regulation by the higher centers of the brain, nor could we survive if individual members or organs of our body rebelled and went their separate ways. Koestler perceives that biological hierarchies function according to "fixed rules and flexible strategies" but does not develop this concept in depth. Polanyi develops this concept in somewhat different language and calls it the principle of boundary control.

MICHAEL POLANYI

Polanyi (1967) writes:

The laws of physics are given in terms of differential equations which determine a definite system only within a set of fixed conditions. Physics is dumb without the gift of boundary conditions, forming its frame; and this frame is not determined by the laws of physics.

For example, we may integrate the classical first-order rate equation

$$-\frac{dN}{dt} = kN \tag{71}$$

where N is the number of particles or molecules present at time t. The rate of decay is proportional to the amount present. Integrating, we have

$$-\log N = kt + C_i \tag{72}$$

where C_i is a constant of integration. Equation (72) represents an infinite number of equations corresponding to the infinite number of possible values of C_i. It is only after we arbitrarily determine C_i by setting boundary conditions such as $N = N_o$ at $t = 0$ that we can use the integrated equation for the quantitative analysis of experimental data.

The boundary conditions often amount to defining an arbitrary scale of measurement. Polanyi continues:

> The laws of chemistry have similar limitations. Generally, to have a definite chemical process, we must frame it by boundary conditions not fixed by the laws of chemistry.
>
> We speak of such boundaries as "fixed conditions" rather than "controlling principles," for their intervention, though indispensable, is not highly significant. This is different for a machine. The boundary conditions of the physical-chemical changes taking place in a machine are the structural and operational principles of the machine. We say therefore that the laws of inanimate nature operate in a machine under the control of operational principles that constitute (or determine) its boundaries. Such a system is clearly under dual control.
>
> Administrative hierarchies are common examples of a higher authority governing lower levels, while relying on the autonomous workings of these lower levels. Hierarchies formed by successive levels of the organism have been described similarly. My own theory expands the structure of hierarchic levels to the relation between biological principles and the laws of physics and chemistry. Biological principles are seen then to control the boundary conditions within which the forces of physics and chemistry carry on the business of life. This dual action of a system is said to work by *the principle of boundary control.*

In his book, *The Tacit Dimension*, Polanyi (1966) frames the principle in evolutionary terms. A lower level of biological organization cannot gain control of its own boundary conditions; but, when a higher level emerges, it does so by fixing the boundary conditions of the lower level. This lower level may then be relied upon to function automatically like a machine, leaving the higher level free to experiment with higher operational principles which may pave the way for the emergence of the next higher level.

Now we must go one step further than Polanyi and ask the questions: "What fixes the boundary conditions?" "What is the nature of the boundary condition?" I propose the hypothesis that the principles of game theory fix the boundary conditions prerequisite to the evolution of higher forms. Koestler's language is prophetic in the way it blends the concepts of game theory with the concept of a hierarchy. I shall show quantitatively that the informational variables of the DNA of higher organisms display the max min and the min max of classical game theory. This is a highly specific quantitative way of fixing the boundary conditions which draw the lines of demarcation between levels of the hierarchy.

GAME THEORY

Game theory is a relatively new branch of mathematics, developed largely by the late John von Neumann. As so many topics in mathematics do, game theory takes words from common everyday English usage and gives them a special technical meaning. A *game* is any conflict situation between two or more opposing sets of interests, and the sets of interests themselves are called *persons* regardless of the number of individuals involved. For example, football is a two-person game.

One of the most common forms of a two-person game is that of a single player against "Nature," some impersonal, antagonistic force. The living system in its struggle for survival is a lone player against "Nature." Therefore we are dealing with a two-person game. A *pure strategy* is a plan of action that is complete. It takes into account any contingency which may arise and has a definite course of action laid out which may be executed regardless of what the enemy does. This plan may be good or bad; the only requirement is that it be complete.

The fundamental concept of game theory is this. If a player wishes to "play it safe" rather than "shoot the moon," there is a way to select a strategy that will maximize his minimum winnings or minimize his maximum losses, regardless of what the enemy does. The best way to illustrate this concept is with an example.

The matrix of Fig. 21 represents a two-person war game. The rows of the matrix represent the strategies of the defensive side, which we will call the "good guys" or G.G., and the columns represent the strategies of the aggressors, the "bad guys" or B.G. The numbers in the matrix represent the number of days between battles. The B.G. want to harass the G.G. by attacking as often as possible, so a greater number of days between battles indicates a loss to the B.G. or a winning to the G.G. Let us examine the matrix more closely.

FIGURE 21

A WAR GAME.
[a]MIN MAX; [b]MAX MIN.

		Bad guys				Row mins
		A′	B′	C′	D′	
	A	7	2	5	1	1
Good guys	B	2	2	3	4	2
	C	5	3	4	4	3[b]
	D	3	2	1	6	1
Column maxs		7	3[a]	5	6	

The G.G. have four strategies, A, B, C, D, and the B.G. also have four, A′, B′, C′ and D′. If the G.G. use plan A and the B.G. plan A′, there will be seven days between battles. However, if the G.G. use plan A and the B.G. plan D′, there will be only one day. Glancing at their other strategies, the G.G. realize that plan D can also possibly yield only a one-day rest, whereas plan B gives two days at the least and plan C guarantees a minimum of a three-day rest. The latter guarantee appeals to them and they decide to use only plan C. They have intuitively utilized the fundamental concept of game theory; for, if we list the minimum value of each row as shown in Fig. 21, the value of three from row C is the maximum of these minimum values. They wish to maximize their minimum winnings. The B.G., on the other hand, wish to attack again as soon as possible. Glancing at their strategies, they see that under plan A′ they could be defeated so badly that it would take seven days to recoup, under plan D′ possibly six days, under plan C′ possibly five days; but plan B′ guarantees that they could attack again in three days at the most. Thus, if we list the maximum values from each column, the value of three from column B′ is the minimum of these maximum values. If the B.G. choose strategy B′, they can minimize their maximum losses.

In the game we have chosen, the max min and the min max

happen to be equal. In such cases, the game is said to have a saddle point. This is not always true, but even in the more complicated situations where it is not it is still possible for a player to maximize his minimum winnings or minimize his maximum losses by choosing a mixture of his pure strategies.

THE MAX MIN AND MIN MAX OF DNA

Now let us reexamine the nature of the boundaries that have been placed on vertebrate DNA. Table 6 lists the maximum and minimum values of D_1 and D_2 for phage, bacteria, vertebrates, and invertebrates. We observe that the vertebrates have minimized the maximum value of D_1. This is another expression of the fact that vertebrate DNA base composition does not vary widely. D_1 is a function only of the p_i, the base composition, and vertebrate base composition does not go to the extremes of the phage and bacteria. As the base composition diverges from the central mean of 50% (C + G), D_1 increases. Obviously, the vertebrates have minimized the maximum value D_1 can take. We also observe that the vertebrates have maximized the minimum value of D_2, the measure of ordering or complexity.

The game theoretic boundaries on the informational properties of vertebrate DNA may be expressed in several equivalent ways. Refer to the entropy scale of Fig. 22. Visualize the following mechanical analogy. The lines, log a, H_1, and H_M, are analogous to solid bars parallel to the horizontal plane. D_1 and D_2 are analogous to flexible springs connecting these bars. This is shown in Fig. 22. A tendency to minimize the maximum value D_1 can take is analogous to a vertical L bar which prevents D_1 from stretching beyond the limit set by the L bar. Similarly a tendency to maximize the minimum value D_2 can take is analogous to an L bar attached to H_1 which prevents D_2 from collapsing beyond its limit. Referring to this mechanical analogy, it is easy to see that the same process which sets a min max on D_1 will also set a max min on H_1. Similarly the max min on D_2 fixes a min max on H_M.

But H_1 and H_M are both entropies. We see here an example of why the second law of thermodynamics is inadequate in describing the evolution of living systems. The "entropy" does not seek a simple maximum; H_1 seeks a max min, and H_M seeks a min max. Part of the divergence, D_1, seeks a min max and the other part, D_2, seeks a max min.

FIGURE 22

MECHANICAL ANALOGY OF GAME
THEORETICAL CONSTRAINTS.

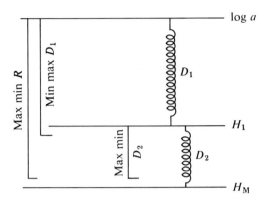

We may regard these variable pairs as the boundary variables at the lowest level of a functional informational hierarchy. The boundary conditions may be described in two equivalent ways:

$$\begin{Bmatrix} \text{min max } D_1 \\ \text{max min } D_2 \end{Bmatrix} \quad \text{or} \quad \begin{Bmatrix} \text{max min } H_1 \\ \text{min max } H_M \end{Bmatrix}$$

We now move to the boundary conditions of the next higher level of the hierarchy. Table 5 shows that vertebrates have a max min of R and a min max of $RD1$. In our mechanical analogy a max min of R would be represented by an L bar attached to the log a bar, preventing the $D_1 + D_2$ system from collapsing below its limit. This is shown in Fig. 22. Here we see the pattern of structure within structure, boundary within boundary, which is the essence of hierarchical organization.

Since

$$R = \frac{D_1 + D_2}{\log a}$$

and

$$RD1 = \frac{D_1}{D_1 + D_2}$$

both R and $RD1$ are functions of both D_1 and D_2, they contain a max min and a min max within them. The lower-level variables, H_1, H_M, D_1, and D_2, take on dimensional values usually expressed in bits, but R and $RD1$ are both dimensionless fractions.

Actually the lowest level of the hierarchy involves only one variable, D_1, which is itself fixed within max min and min max limits. Then D_1 pairs with D_2 to become one of the conjugate variable pairs at the next level of the hierarchy. Next the sum, $D_1 + D_2$, or the information density, becomes a fundamental variable at the next higher level of the hierarchy which now has dimensionless conjugate pairs, R and $RD1$. Thus each advancing level of the hierarchy of game theoretic variables contains within it and is dependent upon variables from the last lower level of the hierarchy.

Since the fundamental objective in game theory is to maximize your minimum winnings and minimize your maximum losses, we must regard all those quantities associated with a max min as winnings or advantages and those associated with a min max as losses or disadvantages. Interestingly enough, this view gives rise to a total conceptual picture which is not only internally consistent but almost predictable from the nature of the information concept.

Consider the max min of R. Redundancy is a capacity to combat error and, if we assume that the successful evolution of vertebrates means that this capacity has been utilized, then maximizing the minimum redundancy is equivalent to minimizing the maximum error, an intuitively reasonable prerequisite to the evolution of higher organisms. The other boundary, a min max of $RD1$ or, equivalently, of D_1, is related to the element of message variety. In Chapter 4 we showed quantitatively that, as the letters diverge from equiprobability and D_1 increases, the potential message variety declines. Hence minimizing the maximum value of D_1 is equivalent to maximizing the minimum potential message variety, an obvious prerequisite to the complexity of higher forms.

In this discussion every quantity associated with a max min is an advantage, and every quantity associated with a min max a disadvantage. However, we must always remember that these are winnings and losses of a particular kind. They are obtained under the guarantee that the winnings will not fall below the max min and the losses will not rise above the min max, and as such they have a somewhat dual character.

In Chapter 3 in the section on language we pointed out that in order to formulate vocabulary and syntax the entropy must decrease somewhat. This is equivalent to defining error. Therefore, although potential message variety is an advantage for winning at the game, this winning must have a little loss mixed in with it before one can formulate language or play the game successfully. Similarly, at the other extreme, the losses have a little winning blended in. If the error is reduced too much, communication is impossible because the message variety vanishes. Thus, although error is a disadvantage, it is advantageous to retain some capacity for it. The pitfalls at the extremes are avoided by the game theoretic boundaries which the vertebrates have placed on their entropy variables. The very nature of the game theoretic concept is to obtain an optimal blend of opposites. The max min allows the winnings to be reduced somewhat but sets boundaries which prevent this process from going too far. Similarly the min max allows the error to remain finite without getting out of hand.

The lower levels of biological organization exemplified by the phage and bacteria have placed no such limitations on the variation of the fundamental informational parameters of their DNA. They let the error become too high and the message variety become too low. They are like the mechanical system of Fig. 22 without the restraining L bars.

In Chapter 4 we noted that the second theorem principle declares its moratorium on the uncertainty restriction only within limited intermediate ranges of the variables. If the system wishes to take advantage of this opportunity, some optimization principle such as that of game theory *must* intervene and place on the situation boundaries that prevent any one of the variables from going to extreme values. Therefore, when the higher organisms emerged, they did so by placing game theoretic boundaries on the entropy variables of their DNA to minimize the maximum error and maximize the minimum variety. In short, in the game of life the higher organisms have selected a successful strategy which has advanced them to a higher level in the hierarchy of living systems.

In reading this chapter the evolutionary thinker surely will have noticed many parallels between the game theoretic concepts and those of classical Darwinian theory. When we say that the living system is a lone player against nature, the evolutionist may regard this as an alternative description of the process of random mutation and natural selection. Organisms which are eliminated in this deadly "game" are poor players. The game theoretic and information theoretic concepts enrich our understanding of this process by explaining *why* some organisms are at a disadvantage at the level of information storage and transmission.

However, these concepts are not merely a rephrasing of classical evolutionary concepts. In Chapter 9, I shall show that they lead us into the consideration of a new evolutionary principle which is essentially the confluence of Darwin's principle of natural selection and Shannon's second theorem. They lead us into the consideration of a new type of selection which I call second theorem selection. I shall show explicitly that this type of selection can distinguish between DNA sequences which are indistinguishable under current Darwinian or non-Darwinian theories.

6

Behind every closed door is a
mystery sealed with seven seals.
—Kahlil Gibran, Sand and Foam

THE GENETIC CODE

When we stated Shannon's second theorem, we noted that "the code is the crux of the matter." All of the promises of the second theorem depend on whether or not the message has been properly encoded at the source. We have witnessed in the decade of the 1960s the discovery of the genetic code, a dramatic and important result. An air of finality and fulfillment pervades the biological world. I do not share this feeling, and I hope this chapter can help to disturb the complacency.

Biologists currently believe that the genetic code is "frozen" or no longer subject to evolutionary change. I am skeptical of this view. Although it is apparent from comparison of protein sequences that the genetic code could not have changed drastically since the emergence of life on this planet, nevertheless it is possible that gradual changes on a broad evolutionary time scale are still occurring. For example, the codon UGA, which incorporates a variety of amino acids with the aid of minor "suppressor" tRNA species, could still be in the process of acquiring a meaning.

We cannot have evolution without variability. The source of the variability which makes possible this "superslow" evolution of the code, which I believe to be possible, is the occurrence of efficient suppressors. These tiny "skeletons in our cupboard" will play a dominant role in this chapter. As Jack Lester King (1971) puts it:

> Actually, the code is rendered slightly ambiguous in the presence of amber and ochre suppressors, and suppressor strains *do* survive.

117

Also, biologists believe that the genetic code is a strict mapping. This concept is intimately interwoven with that of the frozen code. This entire chapter refutes such a view.

It is time to place this important step forward in its proper perspective. It is just that—a first step. The genetic code is a small subroutine of a master program which directs the machinery of life. We have no idea what the language of this master program is like, but we can be sure that it has always evolved, is now evolving, and will continue to evolve in the future.

We should not be surprised and dismayed by the highly redundant DNA now being discovered. I predicted that the rapidly reassociating fractions of Britten and Kohne (1967) would be naturally occurring, highly redundant polymers with a short repeat length (Jukes and Gatlin, 1970). While this paper was in press, Southern (1970) confirmed that the guinea pig satellite DNA fits this description. In 1966, I suggested (Gatlin, 1966) that highly redundant sequences serve control functions in the master program. This prediction is now confirmed by the work of Sivolap and Bonner (1971). As early as 1963, when it was extremely unpopular, I suggested (Gatlin, 1963) that the meaning of even a single codon could be highly dependent on the context in which it appears. We now know this as fact for the START codon.

Therefore, before we even discuss the genetic code, we must first discuss programming principles in order to understand the code's relation to the higher genetic languages which either must exist or are now in the process of evolving.

COMPUTERS AND PROGRAMMING

A computer is a collection of a large number of small electronic components which may be in either of two states. All fundamental electronic components are binary in nature. For example, a switch may be open or closed, a current may be on or off, an iron ring may be magnetized or demagnetized. All numbers and letters may be coded by sequences of the binary digits, 1 and 0, as shown below.

$$\text{ones position} \rightarrow \quad 1 = 1$$
$$\text{twos position} \rightarrow \quad 10 = 2$$
$$11 = 3$$

$$\text{fours position} \rightarrow 100 = 4$$
$$101 = 5$$
$$110 = 6$$
$$\text{etc.}$$

These sequences of 1's and 0's may then be stored in the memory units of computers.

Any computer consists of the following basic units: an input unit which reads the input data and instructions from punched cards or tape; an output unit which prints out the contents of memory cells; an arithmetic unit which performs computations; a memory unit which stores the data in numbered cells; and a control unit which governs the sequential reading and execution of the program. The fundamental difference between a desk calculator and a computer is that a computer can perform a sequence of operations from a list or program which is read into its memory unit.

The control unit then goes into the particular location in the memory unit where the program is stored. It reads the first instruction, executes it, then goes back, reads the second instruction executes it, etc., until every instruction in the program has been sequentially executed. There is one very important exception, a class of statements in a program which are nonexecutable because they do not tell the computer to *do* anything but rather *how* to do something which it must do at some point in the program.

To illustrate, we must introduce an information processing language. There is a hierarchy of language levels. Machine language is a lower-level language most closely related to the mechanical details of information processing within the computer. FORTRAN is the most widely used language at a higher level. With a simple command in FORTRAN, one can call forth a rather complicated series of instructions in machine language. I believe the genetic language has a similar structure.

Let us examine a simple FORTRAN program with a biological objective. This program is listed below.

SUBROUTINE PROTEIN

1 FORMAT (UUU)
2 FORMAT (UUA)

```
  3   IF(CONDITION X. EQ. 1) GO TO 100
      PRINT 1, LEUCINE
      GO TO 101
100   CONTINUE
      PRINT 2, LEUCINE
101   PRINT 1, PHENYLALANINE
      RETURN
```

The first two statements of the subroutine are nonexecutable FORMAT statements. They tell the computer *how* to print. The third statement activates a decision procedure. If a certain condition of the system exists, control is transferred to statement number 100 and leucine is printed out under FORMAT 2, the accepted codon for leucine. Then phenylalanine is printed out under FORMAT 1, its accepted codon. However, if the specified condition does not exist in the cell at that time, the instructions are simply executed sequentially. In this case, leucine would be printed out under FORMAT 1, which is the accepted codon for phenylalanine. Control is then transferred to statement 101 and phenylalanine is printed out under this format. Control then returns to the main program.

I have found this analogy highly repugnant to biologists. Perhaps they are right and I am entirely wrong. I cannot escape the conviction, however, that simply thinking in such terms gives one a completely different perspective on the genetic code, which is valuable. One who thinks in these terms would not refer to the DNA which does not code for protein as "noninformational" DNA, as is currently being done. It is all part of the program even if it is only a CONTINUE statement. Such statements serve extremely valuable programming functions.

I do not mean to imply with the analogy above that I think this type of complicated programming procedure is generally used for the synthesis of proteins. As Arthur Koestler puts it, the living system is a "canon of fixed rules governed by flexible strategies." The genetic code is a canon of fixed rules. The preceding program illustrates how it could be modified in special cases by higher flexible strategies. What this view essentially says is that the genetic code is part of an informational hierarchy. The presently accepted genetic code is a canon of fixed rules which function automatically *in the absence of intervention by a higher level of informational control.*

One of the fundamental operational principles of a hierarchy is that the lower levels must be relied upon to function automatically. The admiral cannot oversee the routine movement of each ship. On the other hand, it must be possible at least to some extent for the higher levels of control to be able to overrule the lower levels, particularly when something goes wrong. Without this principle of overrule, the hierarchy cannot be a goal-seeking or cybernetic system in the same way that the living system is.

Language again presents an analogy. Most words in a language have a fixed meaning which one can look up in a dictionary. However, some words by themselves are ambiguous, capable of being understood in more than one way. When such words are used, their meaning is supplied by the *context* in which they appear. For example, if I use the two phrases, "the root of a tree" and "the root of an equation," there is no question about the meaning of the word, root, in either case. The context has supplied the meaning. We might say that the intrinsic ambiguity of the word, root, has been resolved.

Note that the word, ambiguous, itself has two meanings, its set theoretical meaning which we will define and its common usage meaning. A codon may be ambiguous in the set theoretical sense in that it may give rise to more than one amino acid, yet it may not be ambiguous at all in the common usage sense in that the amino acid inserted at a particular locus may be completely determined by the surrounding context, i.e., the set theoretical ambiguity is contextually resolved.

From the basic principles of language formulation, we can infer that ambiguous words do not make up a large percentage of the vocabulary because the meaning must be supplied by a nonambiguous context. Ambiguous words, far from being a detriment in language, are actually highly useful and even efficient. They prevent undue proliferation of vocabulary by utilizing contextual structure already in existence to determine meaning. If one is limited *a priori* to a dictionary of 64 words of 3 letters each, ambiguity might be invaluable to the evolution of a more sophisticated information processing language.

The development of language begins with a simple vocabulary of completely nonambiguous words. As more complex structures develop, new technical meanings are assigned to the primitive words. For example, the most primitive meaning of the word, root, is the botanical meaning. It is the intervention of the mathematician at a higher level of meaning that has given this word a set theoretical ambiguity.

We must now review the development of the genetic code and present the set theoretic definition of ambiguity before we return to these concepts.

DISCOVERY OF THE GENETIC CODE

The year was 1961. Experimentalists had been studying protein synthesis in the test tube for some time. They knew that, in the supernatant fraction obtained after centrifuging out the debris from a disrupted living cell, there were enzymes, the aminoacyl tRNA-synthetases, which bind a specific amino acid to its specific tRNA. Nirenberg and Matthaei had observed the incorporation of C^{14}-valine into the synthesized protein which was precipitated from the system by hot trichloroacetic acid (TCA-insoluble precipitate). Their system was as follows. From *E. coli:*

Supernatant	Washed Ribosomes	tRNA	Amino Acids
Containing the synthetases and all other unknown factors	Containing the ribosomal RNA and ribosomal protein	Also called soluble RNA	One of which is labeled with C^{14}

Nirenberg and Matthaei (1961) announced that, when they placed synthetic RNA containing only uridine in their system in the presence of C^{14}-phenylalanine, the radioactive counts per minute in the TCA-insoluble precipitate went up from a blank of only 44 to almost 40,000. This change was not due to the high molecular weight of poly-U alone because poly-A and poly-C, which also have high molecular weights, gave no significant incorporation above the blank. The effect was highly specific. Seventeen other amino acids tested gave no such effect with poly-U. Furthermore, the TCA-insoluble precipitate had the physical and chemical characteristics of a synthetic "protein" consisting of a long chain of nothing but phenylalanine residues. After hydrolysis in 12 N HCl followed by electrophoresis, only one spot appeared on the paper, that of C^{14}-phenylalanine.

TABLE 8

ACCEPTED DICTIONARY

$3' \rightarrow 5'$	U		C		A		G	
U	UUU	Phe	UCU	Ser	UAU	Tyr	UGU	Cys
	C	Phe	C	Ser	C	Tyr	C	Cys
	A	Leu	A	Ser	A	STOP	A	STOP
	G	Leu	G	Ser	G	STOP	G	Trp
C	CUU	Leu	CCU	Pro	CAU	His	CGU	Arg
	C	Leu	C	Pro	C	His	C	Arg
	A	Leu	A	Pro	A	Gln	A	Arg
	G	Leu	G	Pro	G	Gln	G	Arg
A	AUU	Ile	ACU	Thr	AAU	Asn	AGU	Ser
	C	Ile	C	Thr	C	Asn	C	Ser
	A	Ile	A	Thr	A	Lys	A	Arg
	G	Met	G	Thr	G	Lys	G	Arg
G	GUU	Val	GCU	Ala	GAU	Asp	GGU	Gly
	C	Val	C	Ala	C	Asp	C	Gly
	A	Val	A	Ala	A	Glu	A	Gly
	G	Val	G	Ala	G	Glu	G	Gly

The conclusion from the experiment was clear-cut and dramatic. A "messenger" RNA containing only uridine had specifically directed the synthesis of a "protein" containing only phenylalanine. Thus, according to current theory, the triplet UUU "coded" for the amino acid, phenylalanine. The first "word" in the genetic code was known.

More definitive experimental techniques were developed by the groups of Nirenberg, Khorana, and others, and within five years the entire genetic code had been worked out; it is shown in Table 8. A more detailed discussion of these experiments may be found in Jukes (1966) or Yčas (1969).

In order to understand the deep excitement which the discovery of the genetic code caused in the scientific community, one must be aware of the years of thought which had preceded it. A background of experimental work had told us that all the chemical reactions taking place in living system are catalyzed by enzymes, all enzymes are proteins, and all proteins are unique sequences made up of twenty kinds of amino acids. After we discovered that DNA is a sequence of four kinds of bases and that it carries the hereditary information, the hypothesis that the base sequence in DNA determines the amino acid sequence in protein presented itself so naturally to the entire biological community that it cannot be attributed to any one individual. How a sequence of four different things can determine or code for a sequence of twenty different things became known as the coding problem. Again the reasoning was instinctive. When confronted with two sets of things between which we know there is some sort of connection, the human mind almost invariably attempts to set up a one-to-one correspondence between the members of the two sets. The mental process that occurred is as follows. One DNA base cannot determine one amino acid in protein; otherwise, the latter would not be uniquely determined. If a sequence of two DNA bases codes for one amino acid, there are 4×4 or 16 DNA doublets, and this is still short of the 20 amino acids. However, if a triplet of DNA bases codes for 1 amino acid, there are $4 \times 4 \times 4$ or 64 possible triplets and this is more than enough to determine each amino acid uniquely.

Thus the idea arose that a sequence of three bases in DNA codes for one amino acid in protein, and during the years when the genetic code was being worked out the groups of Crick (1961) and Khorana (Nishimura et al., 1965) presented convincing experimental evidence that the genetic message is read in units of 3. This unit is now called a codon.

THE MAPPING ASSUMPTION

Let us examine in some detail the precise mathematical basis of the reasoning process above. It is so instinctive that it may appear to the nonmathematician as almost too simple for mathematical analysis. Yet this is precisely the kind of analysis that forms the basis of a large part of modern mathematics and leads to some of its most profound results. For example, the *Principia Mathematica* of Russell and Whitehead (1910) deals with the underlying basis of arithmetic, i.e., statements like $1 + 1 = 2$.

We have already defined a set as a collection of individual entities where the identity of the individual is the dominant idea. Hence the order in which the individuals are listed is immaterial and there is no point in naming the same individual more than once. We need some further definitions.

An *n*-tuple is a sequence of symbols where the order in which the symbols are placed is vital to the identity of the *n*-tuple and any symbol may be repeated; for example;

$$\langle 1, 2, 3 \rangle, \quad \langle 1, 3, 2 \rangle, \quad \langle 1, 1, 2, 3 \rangle$$

are all different *n*-tuples.

A relation is a set whose elements are *n*-tuples. For example, the preceding *n*-tuples could all be regarded as elements of a set which we would list as follows.

$$\{\langle 1, 2, 3 \rangle, \quad \langle 1, 3, 2 \rangle, \quad \langle 1, 1, 2, 3 \rangle\}$$

A *binary relation* is a set whose elements are 2-tuples. Below are three binary relations which differ in important ways.

$$R_1 = \{\langle A, 1 \rangle, \quad \langle B, 2 \rangle, \quad \langle C, 3 \rangle, \quad \langle D, 4 \rangle\}$$
$$R_2 = \{\langle A, 1 \rangle, \quad \langle B, 1 \rangle, \quad \langle C, 1 \rangle, \quad \langle D, 4 \rangle\}$$
$$R_3 = \{\langle A, 1 \rangle, \quad \langle A, 2 \rangle, \quad \langle B, 2 \rangle, \quad \langle C, 3 \rangle\}$$

The *domain* of a binary relation is the set whose elements are the symbols that appear in the first position of each 2-tuple. The domains of relations R_1, R_2, and R_3 are, respectively,

$$DR_1 = \{A, B, C, D\}$$
$$DR_2 = \{A, B, C, D\}$$
$$DR_3 = \{A, B, C\}$$

The *counterdomain* is the set of symbols which appear in the second position of the 2-tuple. The counterdomains of relations R_1, R_2, and R_3 are, respectively,

$$CDR_1 = \{1, 2, 3, 4\}$$
$$CDR_2 = \{1, 4\}$$
$$CDR_3 = \{1, 2, 3\}$$

We are now ready to define one of the most important concepts underlying the genetic code. A *function* is a binary relation such that for every element in the domain there corresponds one and only one element in the counterdomain. This means simply that, given any element in the domain, there is only one element in the counterdomain one can associate with it. Thus relations R_1 and R_2 are both functions, but relation R_3 is not because, given the element A from R_3, either 1 or 2 could be associated with it. Neither one is uniquely determined. Thus the concept of a function has embedded in it the idea of a unique, directional correspondence. The counterdomain of a function is sometimes called its *range*, and the function is said to *map* the domain onto the range. The function itself is often referred to as a *mapping*. A mapping is also called a *code*.

It is clear that, when confronted with the facts of protein structure and function along with those of DNA structure and function, the human mind automatically assumed that a mapping exists. The domain of the mapping is the set of all 64 possible triplet sequences of the 4 DNA bases (or RNA bases), and the range of the mapping is the set of all 20 amino acids found in protein. The mapping which has been experimentally determined for RNA is as follows:

$$\left\{ \begin{array}{l} \langle \text{UUU, phe} \rangle, \quad \langle \text{UUA, leu} \rangle, \quad \text{etc., where the entire mapping} \\ \qquad\qquad\qquad\qquad\qquad\quad \text{can be read from Table 8} \end{array} \right\}$$

Table 8 is now known as the genetic code, and we will refer to it often as the accepted dictionary. It means that, whenever a given triplet of DNA bases is transcribed into messenger RNA for the synthesis of protein, this triplet will give rise to one and only one amino acid in protein. For example, UUU gives rise to phenylalanine only, UUA gives rise to leucine only, etc. This is the *mapping assumption*.

DEGENERACY AND AMBIGUITY

We can now define in more precise terms the differences between relations R_1, R_2, and R_3. Relation R_1 is a one-to-one mapping. See the diagram in Fig. 23. For every element in the domain there corresponds only one element in the counterdomain, *and* for every element in the counterdomain there corresponds only one element in the domain. The latter is not a necessary condition for the existence of a function; but, whenever it occurs, we have a one-to-one mapping.

Two terms have arisen in the literature of the genetic code which we can now define in precise, mathematical terms. The first is degeneracy. Whenever more than one element in the domain corresponds to the same element in the counterdomain, the code is said to be degenerate. This is a many-to-one correspondence and is diagrammed in Fig. 23.

FIGURE 23

TYPES OF RELATIONS.

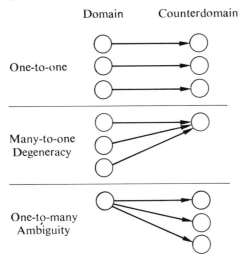

A glance at Table 8 shows that the genetic code is extensively degenerate and there is an underlying pattern to this degeneracy. In eight cases the last letter of the triplet may be any one of the four bases. The initial doublet carries complete information for the specification of a particular amino acid. In thirteen cases the third letter may be either of two bases. In only two cases is there a one-to-one correspondence between codon and amino acid.

This pattern of degeneracy strongly suggests in itself that the present triplet code has evolved from a more primitive doublet code. Jukes (1966) presents this theory in some detail along with a number of convincing arguments in its support.

The second word which has been used extensively in connection with the genetic code is ambiguity. When one element in the domain corresponds to more than one element in the counterdomain, ambiguity exists. This is the one-to-many relation of Fig. 23. When ambiguity occurs in a binary relation, the relation is no longer a function. Relation R_2 is degenerate and relation R_3 is ambiguous and degenerate.

The terms ambiguity and degeneracy are relative terms. They depend upon our definition of domain and counterdomain. A one-to-one mapping, of course, maps in either direction. However, if we reverse the definitions of domain and counterdomain, a degenerate mapping becomes an ambiguous relation and conversely an ambiguous relation becomes a degenerate mapping. If we have a relation that is both degenerate and ambiguous, interchanging the definitions of domain and counterdomain leaves the same type of relation, except that the correspondences which originally produced the degeneracy now produce the ambiguity and conversely the correspondences which originally produced ambiguity now produce degeneracy. Hence one should always define the domain and counterdomain of a relation before using the terms ambiguity and degeneracy. Unless otherwise stated, we will always consider the domain of the genetic code to be the set of 64 triplets and the counterdomain the set of 20 amino acids.

THE CENTRAL DOCTRINE

It seems to be the nature of doctrine, be it religious or scientific, that certain things are strictly forbidden. The central doctrine of the genetic code is the mapping assumption, which is equivalent to saying that ambiguity *in vivo* is strictly forbidden. In 1966, Crick (1966b) made the following statement:

> The skeleton in our cupboard is the possibility of ambiguity. It may be that even in the middle of a message a certain triplet may stand for more than one amino acid. Of course, we know that this can happen when mistakes occur in the translation mechanism, either due to so-called suppressor genes or due to the influence of various small molecules, such as certain antibiotics. What is at issue, however, is whether in a normal cell there is any ambiguity of translation.

In general, biologists believe that there is not. They believe that, if a codon on the messenger RNA were to give rise to an amino acid other than that in the accepted dictionary, this would certainly be a mistake, a rare event, and definitely not a planned part of the normal operation of the organism's information processing machine. In other words, biologists, by and large, do not believe in the type of programmed ambiguity which we illustrated at the outset of this chapter. We already know that the initiation and probably also the termination of protein synthesis are not under the control of a mapping; but most biologists still believe that between the START and STOP signals there is no deviation from the mapping assumption.

But, if the possibility of *in vivo* ambiguity in the "normal" organism can elicit such negative feelings as one would have toward a "skeleton in our cupboard," then the mapping assumption has indeed become holy doctrine; and I have always been suspicious of doctrine. I shall present the heresy that the genetic code is a relation rather than a function. Since I view the genetic code as a small part of a master program which structures the information processing system as an ascending functional hierarchy of controls, ambiguity is no specter in the closet in such a scheme but rather a useful member of the household. Game theoretic principles govern the emergence of higher levels of control and organization, whereas programming functions oversee the routine operation of each level. The mapping function is at the lowest level of this functional hierarchy.

Ambiguity is ubiquitous *in vitro*. When an observation arises that makes us uncomfortable, one of the first results is a voluminous literature describing the phenomenon in exact detail. I will not attempt a review of the literature on "miscoding," but I have listed beside each codon in Table 9, which is an "ambiguous dictionary," some of the amino acids other than those in the accepted dictionary which that codon has been observed to code for *in vitro*.

AMBIGUITY *in vivo* OR SUPPRESSION

The language of science sometimes develops along strange lines. The codons which signal the termination of a particular message should be called STOP codons or END codons, but they have come to be called nonsense codons. There are three of them, UAA, UAG, and UGA. Perhaps the misnomer arose from the idea that they do not specify any particular amino acid, but even this occurs in a phenomenon also misnamed as suppression.

When a mutant phenotype is restored to its original or "wild-type" state by the action of a second mutation at a different site on the genome, the second mutation is said to "suppress" the first. The phenomenon of suppression has been studied extensively in relation to the three

TABLE 9

AMBIGUOUS DICTIONARY[a]

	U	C	A	G
U	UUU leu,[1] ileu[1]	UCU	UAU	UGU trp,[2] val[2]
	C	C	C	C
	A	A his[2]	A lys[2]	A arg,[2] asn,[2] asp,[2] cys,[2] gln,[2] glu[2]
	G	G	G asp[2]	G gly[2]
C	CUU	CCU	CAU tyr[2]	CGU cys,[2] val[2]
	C	C	C thr[2]	C
	A asp,[2] gln,[2] lys[3]	A his[2]	A	A
	G	G arg[2]	G asp[2]	G gly[2]
A	AUU	ACU	AAU lys[2]	AGU cys[3]
	C	C	C thr[2]	C ala,[2] cys[3]
	A met[2]	A	A	A
	G	G	G	G gly,[2] phe,[2] trp[2]
G	GUU	GCU	GAU	GGU
	C	C	C	C arg[2]
	A	A	A	A glu[2]
	G trp[2]	G arg[2]	G asp[2]	G asp,[2] trp[2]

[a] References: 1. Davies *et al.* (1964). 2. Soll *et al.* (1965). 3. Brimacombe *et al.* (1965).

nonsense codons and is termed nonsense suppression. A first mutation occurs in the genotype of a bacteriophage producing a STOP codon. The synthesis of the protein in which the mutation occurs is cut off in the middle of the message, and the resulting, incomplete protein fragments have been isolated. Naturally, these mutants do not grow well on most strains of their bacterial host, *E. coli*. However, there are some *E. coli* strains called suppressor positive (su⁺) strains which permit the growth of these mutant phage to varying degrees. Sometimes the growth rate is restored to that of the normal wild-type phage after transfer to the su⁺ strain. In such cases we could say that the suppression is 100% efficient. The mechanism of this suppression has been shown to be the presence in the bacterial host of a special tRNA species which is capable of translating the nonsense codon as an amino acid. Jukes (1966) gives a more detailed description of this entire process.

Most suppressors observed to date are very inefficient. However, it is extremely significant that 100% efficient suppression has been observed at all. It does prove that suppressors *can* restore completely normal function. If a mutant were to arise in a host already carrying an efficient suppressor tRNA species, it would be passed over as wild type and it would be the type most likely to survive in nature. The inefficient suppressors are the ones most likely to be picked up in the laboratory.

An interesting study by Garen and Siddiqi (1962) confirms this analysis. They studied *E. coli* mutants unable to synthesize the enzyme alkaline phosphatase, except in certain su$^+$ strains. They measured *quantitatively* the amount of enzyme synthesis restored relative to wild type. These values ranged from 3 to 100%. They state:

> It is striking that one of the P$^-$ mutants enables the fully normal amount of enzyme to be synthesized, thus completely restoring the function of the P cistron despite the continued presence of the P$^-$ mutation. This mutation would not have been detectable if the screening for P$^-$ mutants had been carried out with a strain that contained the suppressor.

Thus I consider it extremely significant that we have observed even one case of 100% efficient suppression. It implies that only the mutant codon, and possibly other codons at sites which do not affect function, have been translated ambiguously.

Throughout this discussion it is assumed that a mutation has occurred in the anticodon of the su$^+$ tRNA species which allows it to pair with the STOP codon. This is naturally what one would assume under the currently accepted coding concepts. However, in at least one case we find that this assumption is incorrect. The tryptophan tRNAs for both the su$^-$ and su$^+$ strains of *E. coli* CAJ64 have been sequenced (Hirsh, 1970) and both have the same anticodon sequence, $\overrightarrow{CCA}$.

According to all the currently accepted rules, this anticodon should pair only with the codon $\overrightarrow{UGG}$, the accepted codon for tryptophan. However, both tRNA species recognize $\overrightarrow{UGA}$. The su$^+$ species does this more efficiently, yet the only difference between the two sequences is at position 24, which contains an adenine in the su$^+$ strain in place of a guanine. There is no evidence for any other tryptophan tRNA in the su$^+$ strain.

Here we are observing the influence of "context" on the meaning of a codon. In this case it involves the entire conformation of the tRNA species. Even a small change at a site distal to the anticodon can influence codon meaning. Even one with such radical ideas as I would never have expected such a fine degree of influence from the surrounding context.

In addition to nonsense suppressors, there are missense suppressors which translate codons in ways other than those prescribed by the accepted dictionary. For example, suppressors have been observed which translate the aspartic acid codon as glycine (Berger and Yanofsky, 1967) and the arginine codon as glycine (Carbon et al., 1969). These are without question observations of the occurrence of ambiguity in vivo. However, biologists believe that these observations in microorganisms do not represent the normal state of affairs, especially in higher organisms. They are an anomaly, a spurious peak on the scope of reality, and not to be taken too seriously. The same attitude was the historical reaction to the first experimental observations which suggested that the hereditary material was DNA, not protein, as everyone had believed.

Hence, under current doctrine, it is strictly forbidden that suppressor tRNA species, these small skeletons in our cupboard, exist and function in normal organisms. Of course, it is an artificial restriction to state the central doctrine only in terms of suppressor tRNA species because we know that ribosomes can also play an active role in miscoding. Although this is not at all understood as yet, we do know that the protein components of the ribosome play a vital role in determining the specificity of codon recognition (Traub and Nomura, 1968).

One protein must be present to facilitate miscoding by various agents (Ozaki et al., 1969), and another protein must be present to prevent a high frequency of spontaneous miscoding (Nomura et al., 1969). Hence ribosomal proteins also play a role in the frequency of translational errors.

Rapidly accumulating data indicate that ribosomes are chemical individuals. There are classes of ribosomal protein present in amounts corresponding to less than one copy per ribosome (Nomura et al., 1969). Kurland (1970) in his review of these data makes the following statement:

The overwhelming impression is that the ribosome is far from being a primitive organelle. Instead, it seems to be a highly evolved and complicated entity, containing a much larger number of components than can be accounted for by present views of protein synthesis.

Under these present views of protein synthesis, one is faced with the following dilemma in regard to missense suppression (also called extragenic suppression). Let us suppose that a given codon A is assigned to amino acid α in the accepted dictionary. A mutation occurs producing codon A, and a suppressor tRNA species arises which translates codon A as amino acid β. Under the mapping assumption, the suppressor tRNA would be just as likely to translate codon A at other positions in the messenger and even in the messengers of other proteins as at the mutant site. According to current doctrine, this sudden sweeping change in all the proteins of an organism would be lethal.

Indeed it is difficult to devise a consistent explanation of the viability of missense suppressors under current doctrine. The attempted explanation runs as follows. A few protein molecules with the ambiguous translation and hence partially restored function are produced, and this is sufficient to make the organism viable at a reduced growth rate. Therefore, if the organism is viable at all, the explanation implies that the suppressor tRNA has translated only at the mutant site or at a highly restricted set of sites, and all other codon A sites in the organism's proteins have been translated normally. Yet such site selective translation is quite contradictory to current coding concepts.

Undoubtedly, the ribosome plays a vital role in this selective process. For example, mutants of *E. coli* are known which are called "conditionally streptomycin-dependent" or CSD mutants (Gorini and Kataja, 1964). One such mutant cannot synthesize arginine, hence will not grow unless arginine is supplied in its growth medium. However, the arginine auxotroph can grow without arginine if streptomycin instead is supplied in the growth medium. This phenomenon is called "phenotypic repair" and is known to be due to a suppressor mutation which alters the structure of a 30S ribosomal protein such that "miscoding" occurs in the presence of streptomycin (Gorini *et al.*, 1966).

Miscoding due to the action of streptomycin has been extensively studied *in vitro*. It is generally agreed that, if anything remotely approaching the degree of miscoding observed in the test tube were to occur *in vivo*, this drastic change in the organism's proteins would be lethal. Hence we are forced to conclude that in the viable organisms exhibiting phenotypic repair the ambiguous translation of codons has been selectively screened and extensively restricted by the ribosome.

The ribosome should be regarded more as a master craftsman than as a robot at a printing press. As the messenger RNA moves through the interior of the ribosome during protein synthesis, a portion of its sequence is "engulfed" in the ribosomal structure and becomes resistant to ribonuclease, an enzyme which hydrolyzes unprotected RNA (Takanami *et al.*, 1965). Also about 30 to 35 amino acid residues in the nascent polypeptide chain are protected from proteases (Malkin and Rich, 1967). Thus the "fingers" of the ribosome reach out to enclose a substantial portion of the input sequence and the output sequence of the channel.

I suggest that the ribosome has the ability to "scan" or "search" base sequences in its interior and "select" sequences which are admissible to the information processing system and reject those which are not. As the higher organisms have evolved, this decision procedure has become more efficient.

Such a mechanism would give the system the ability to read a codon in the context of a sequence of codons and the entire translational environment. It is quite clear already that this is precisely what it does in the case of the AUG codon, which is instrumental in initiating protein synthesis.

The protein molecules now called initiation factors, which associate transiently with the ribosome, also play a role in this recognition process. There are also protein molecules called σ or γ factors (Burgess *et al.*, 1969; Krakow *et al.*, 1969) which are instrumental in initiating the transcription of DNA by RNA polymerase.

Perhaps there are "translation factors" similar to the initiation, termination, and σ factors which temporarily alter the topography of the ribosome such that a given codon, like the mutant codon in a viable missense suppressor, is preferentially translated to an amino acid other than that in the accepted dictionary. This entire process could well become a programmed, reproducible operation upon which the survival of the organism depends. Such a view would replace the current doctrine that a unique *codon* in DNA always gives rise to the same *amino acid* in protein with the concept that a unique sequence of bases in DNA always gives rise to the same unique *sequence* of amino acids in protein within a given biological context. Such a view would resolve the suppression dilemmas.

UNIVERSALITY OF THE CODE

The preceding discussion is not intended to be a detailed proposal of the design and function of the living computer's hardware. The purpose of this book is to present new and useful operational principles of information processing in the living system. If these principles are true, experimentalists will discover the hardware to implement them. However, it is ironically true that experimentalists can discover hardware whose function they do not understand because it does not fit into the current theory of function.

It is also quite true, as the recent discovery of RNA-dependent DNA polymerase (Temin and Mizutani, 1970) so dramatically illustrates, that we do not see rather easily demonstrated components of the living computer's hardware simply because we do not believe they are there and hence have never looked for them. In a system as complex as the living system, there is no point in worshipping the "hard" experimental fact, because our theoretical perspectives have played no small role in determining which facts we know. And, even if we were to describe completely every component of the living computer's hardware, we are still left with Polanyi's concept that one cannot understand the functioning of a complex machine from a description of its hardware alone, no matter how complete this description may be.

This is why, every so often, we discover a small piece of hardware whose very existence makes biologists uncomfortable. A case in point is the small suppressor tRNA species. Take, for example, the following statement by Yčas (1969):

> ... since it is now known that "suppressor" mutations can change a codon reading, universality can no longer be taken for granted; indeed suppression suggests that it is universality

which is unexpected. Nevertheless, the code seems to be the same in all organisms.

In a short note, entitled "Translation Not Universal," by an unidentified correspondent of *Nature*, Volume 227, page 999 (1970), the work of Hunter and Jackson (1970) is reviewed. They carefully examined the amino acid sequences of rabbit hemoglobin synthesized *in vitro* with *E. coli* aminoacyl tRNAs. As the correspondent puts it, they "sadly conclude that substantial levels of miscoding occur during translation in this heterologous system."

Hunter and Jackson find five cases in which cysteine replaces arginine and several cases where methionine was being introduced in place of valine. The correspondent refers to these findings as "somewhat traumatic results." The results of the Michelson-Morley experiment were also "traumatic" to the physicists of that day because they believed in the existence of the ether. If experimental results such as those above are "traumatic" to biologists, it is because they believe in the holy inviolability of the central doctrine.

I believe that cases will be found, particularly in higher organisms, where a given codon is translated as an amino acid other than that in the accepted dictionary, and that this "miscoding" is a programmed, reproducible part of the normal functioning of the information processing system.

Under such a view, the genetic code may be regarded as "universal" in the sense that all organisms use the same dictionary as a canon of fixed rules but is not "universal" in the sense that all use the same pattern of overrule whereby the meaning of a given codon may be modified by the context in which it appears. In other words, all organisms do not have the same program even though the programs may all be written in the same information processing language.

INITIATION AND TERMINATION

All programs must have some sort of a START and STOP signal or instruction. It was assumed at first by biologists that the initiation and termination of polypeptide synthesis was a simple mapping function. They believed that the codon AUG (and also possibly GUG) was the START signal, that any one of the three codons, UAA,

UAG, or UGA, was the STOP signal, and that no other signals were necessary. We know now that this is an oversimplified view.

The codon GUG binds *in vitro* to formylmethionine tRNA, whose anticodon is CAU. These codons are always written in the *internal* $3' \rightarrow 5'$ direction of RNA. Since the codon and anticodon align in antiparallel directions for hydrogen bonding, just like the complementary strands of DNA, we have the following situation:

$$3' \rightarrow 5'$$

Codon	GUG on mRNA
Anticodon	UAC on tRNA

$$5' \leftarrow 3'$$

at alignment for polypeptide synthesis.

A sequence of alternating GUGU . . . gives rise *in vitro* to f-met-cys-val-cys-val . . . (Ghosh *et al.*, 1967). Because of this experiment *in vitro*, many biologists now believe that GUG may bind to formylmethionine *in vivo* and hence initiate protein synthesis. If this is true, the codon GUG is ambiguous in the set theoretic sense. Therefore, if the GUG-CAU association does occur *in vivo*, this is a deviation from the mapping assumption.

The initiation and termination functions are programming functions because they are higher-level informational functions too important to be left under the control of a mapping. This means that the lower-level informational structures do not contain sufficient information to perform this more sophisticated function. There is now ample experimental evidence that the START and STOP codons possess *necessary but not sufficient* information to perform their functions of normal chain initiation and termination *in vivo*.

The biological context which initiates polypeptide synthesis is extremely complex. At the present time we know that it involves at least three factors, F_1, F_2, and F_3, magnesium ion, and guanosine triphosphate. Within this biological context the 30S ribosome attaches itself to an initiation site on the mRNA. The current expectation is that we will find some specific sequence or secondary structure which will identify this site. However, although we already know the base sequence at the start of six genes, the molecular explanation of initiation site recognition remains obscure. Next formylmethionine tRNA becomes attached to the 30S ribosome, which is then joined by a 50S ribosome to form the 70S ribosome which reads the message until the termination signal is reached, whereupon the complex dissociates.

There are two tRNA species for methionine that have been completely sequenced and are quite different in their primary structures. One becomes formylated in prokaryotic, but not in eucaryotic organisms and binds only to an AUG codon that initiates synthesis. The other does not become formylated. It binds only to 70S ribosomes, and hence recognizes AUG codons only in the interior of a polypeptide message.

The AUG codon appears in an untranslated region of the messenger RNA of phage R17 (Steitz, 1969), where it does not initiate normal protein synthesis because it is out of phase. If this RNA is treated with formaldehyde, larger quantities of the dipeptides which occur at the start of each phage protein are produced and some new ones appear. The interpretation is that the "internal" AUG codons which do not initiate protein synthesis *in vivo* do so in this artificial test tube environment (Lodish and Robertson, 1969). This experiment confirms the concept that the living organism possesses such a highly structured informational context that the same word, AUG, has different meanings in different contexts. When it appears in an initiation site, it means START, and when it appears in the interior of a polypeptide message, it means methionine.

Normal chain termination is also a complex process requiring at least two release factors, R1 and R2. Also a protein factor named ρ-factor has been found which causes specific, nonrandom termination of RNA synthesis (Roberts, 1969). New factors involved in protein synthesis are still being discovered, and every addition to this list only increases the complexity of the context in which the genetic code is interpreted.

If an amber codon arises in the interior of a protein message by mutation, this codon is sufficient to terminate polypeptide synthesis, at least in some cases. However, this termination process is different from the normal chain termination process (Webster and Zinder, 1969).

Nichols (1970) has found an RNA fragment from R17 which contains the proper codons for the last six amino acids in the phage coat protein. The sequence terminates with *two* successive STOP codons, UAAUAG.

It was at first thought that perhaps tandem STOP codons were generally used in termination. However evidence rapidly appeared that only one terminator codon occurs in some cases (Rechler and Martin, 1970; Clegg *et al.*, 1971).

All of this experimental evidence supports the concept that the START and STOP codons contain necessary but not sufficient information to perform their function of normal chain initiation and termination *in vivo*. With this expanding wealth of experimental knowledge regarding the complexity of the normal chain initiation and termination processes, biologists are beginning to appreciate the subtleties of the context in which the START and STOP codons function. I feel that there is no reason to believe that such subtleties of context could never play a role in the meaning of a codon in the interior of the protein message.

In fact, if we do not adopt this concept, then we are again faced with a dilemma in regard to nonsense suppression analogous to the one we described for missense suppression. If one of the three STOP codons, UAA, UAG, or UGA, is sufficient for normal chain termination, then this particular codon must not occur anywhere within a protein message. If a mutation occurs which places this codon in the middle of a message, incomplete protein fragments are formed. This mutation is then suppressed by a tRNA species which can translate the mutant STOP codon as an amino acid. If the mutant STOP codon and the normal chain-terminating codon were the same, under the mapping assumption the mutant tRNA species would also translate the normal STOP codon as an amino acid and the cell's proteins would be linked together continuously. Presumably this would be lethal to the cell.

Biologists are already well aware of this dilemma and have tried to avoid it as follows. The amber codon, UAG, and the codon, UGA, are readily suppressed, i.e., the mutant phages grow well in many of their respective suppressor hosts, some as well as wild type. It was at first the observation that the ochre codon UAA is not so readily suppressed and grows very slowly in its suppressor host. It was theorized that UAA is the normal chain-terminating codon in *E. coli* and the slow growth of ochre suppressors is due to the frequent translation of normal STOP codons by the suppressor tRNA. However, there are reports of the observation of 100% efficient ochre suppressors (Gartner *et al.*, 1969), and the mechanism of this 100% efficient ochre suppression has been shown to be the same as any other, namely, the presence of a minor tRNA species.

The obvious way out of this dilemma is the simple principle that a single STOP codon contains *necessary but not sufficient* information to perform the function of normal chain termination *in vivo*. This is simply another way of saying that the START and STOP functions are not under the control of a mapping. They are functions programmed by a higher level of informational authority which can supply additional information to the lower level of function. Thus our original assumption of a direct mapping from the START and STOP codons is incorrect.

Another mapping assumption which has proved to be incorrect involves the tRNA molecule. It was at first assumed that there was a one-to-one mapping from a codon to its tRNA species. We may think of this mapping in terms of submappings for each amino acid. The domain consists of all the codons for that particular amino acid, and the counterdomain is the set of all tRNA species which bind that amino acid. We now know of cases in which there are more tRNA species than there are codons for a given amino acid. For example, there are four tRNA species in brewer's yeast for lysine, which has only two codons (Bergquist, 1966) and three tRNA species in *E. coli* for tryptophan, which has only one codon (Muench and Berg, 1966). In such cases, at least one of the codons must correspond to more than one tRNA species. This is ambiguity under the definition of domain and counterdomain given above. It is also known that some tRNA species bind to more than one codon. This is degeneracy, and it is the situation which gave rise to the wobble theory (Crick, 1966a) which for the first time suggested that base pairs other than the standard $C \cdot G$ or $A \cdot T$ pairs exist between the third base of the codon and the first base of the anticodon.

We should recall here that, although ambiguity and degeneracy are relative terms, when we have a relation that is both ambiguous and degenerate neither of these properties can be erased by reversing the definition of domain and counterdomain. Thus, although current theory assumes a strict mapping from codon to amino acid, no mapping of any kind exists between codon and tRNA, the direct linkage between the two, no matter how we define domain and counterdomain. In fact, an unexpected variety and proliferation of tRNA is now being observed throughout nature. I strongly recommend to the reader a fascinating article by K. Bruce Jacobson (1971) in which he discusses the "strange" tRNA species—"strange" because they do not fit into the current conceptual framework. The situation is reminiscent of the physicists and their "strange" particles.

Thus two mapping assumptions regarding the genetic code have come and gone. Therefore I will continue to be skeptical of the central doctrine until all the evidence is in. To map is a compulsive tendency of the human mind and, if relativity has taught us anything, it is to be suspicious of these "common sense" forms of reasoning. They do not necessarily correspond to physical reality, and yet we are so drawn to them that it takes a disproportionately large amount of experimental evidence, in fact virtually conclusive experimental evidence, to displace them.

The discovery of the genetic code was a magnificent achievement, but even more fascinating developments are imminent.

7 *It would seem to me that man cannot live without mysteries.*
—Erwin Chargaff (1971)

HIGHLY REDUNDANT DNA

If the genetic code and the assembly of protein sequences are sub-routines of a master program, we must inquire into the nature of this master program. What are its language and structure? We have already mentioned that it must have a hierarchical structure with higher-level programs generating and controlling each lower-level program. In this chapter we will make a first attempt to describe in fundamental terms the nature of some of the sequences in the master program language which perform control functions.

In Chapter 6, I noted that theory can determine which experimental fact is known. In all fairness, it is also true that experiment can guide and stimulate theoretical development. This is certainly true in the case of information theory. Therefore in this chapter I shall present first certain recent experimental observations which, I feel, will be influential in extending theory.

THE EXPERIMENTS OF BRITTEN AND KOHNE

In the mass of experimental work that our technology has made possible today there appears occasionally an experiment of singular beauty and significance. In my opinion, the work of Britten and Kohne (1967) is of this nature. We will review the experimental details.

Double-stranded DNA molecules were sheared to an average length of about 400 nucleotides and denatured (separated into single strands). Then under very carefully controlled conditions these relatively short single strands were allowed to reassociate. This is clearly a bimolecular reaction, and the rate data follow the expected second-order kinetics. For an ideal second-order reaction,

$$-\frac{dc}{dt} = kc^2 \tag{73}$$

where c is the concentration of reacting species, t is the time, and k is the second-order rate constant. Integrating under the boundary condition that $c = c_0$ at $t = 0$, we have

$$\frac{c_0}{c} = kc_0t + 1 \tag{74}$$

When $c = c_0/2$ and $t = \tau$, the half-life or time for half of the reaction to be completed, then

$$k = \frac{1}{c_0\tau} \tag{75}$$

Britten and Kohne report their rate data as a plot of the fraction reassociated $(1 - c/c_0)$ versus the log of c_0t. It is shown in Fig. 24.

FIGURE 24

RATE DATA OF BRITTEN AND KOHNE.

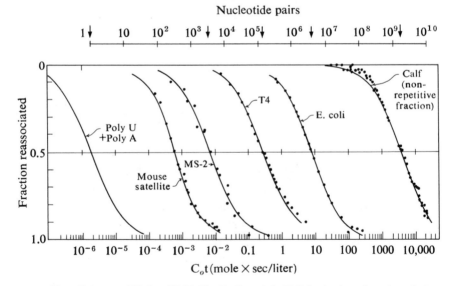

From Britten and Kohne (1968, Fig. 2). Copyright 1968 by the American Association for the Advancement of Science.

$c_0\tau$ and hence k may be obtained from this plot. Britten and Kohne always discuss rates in terms of $c_0\tau$ rather than k. Since k and $c_0\tau$ are reciprocals, a faster rate means a smaller $c_0\tau$.

Britten and Kohne observed that the DNA of higher organisms contains fractions which reassociate at fantastically greater rates than the main fraction. For example, calf thymus DNA has a fraction comprising about 38% of the total DNA which reassociates about 10^5 times faster, and one comprising 3% which reassociates about 10^6 times faster than the main DNA. The rapidly reassociating DNA appears to be universally present in all higher organisms above bacteria, different organisms showing different "spectra" of rapidly reassociating species as shown in Fig. 25. There is no evidence of variation between different tissues of the same species or even of the same organism.

Previous work had shown that, the larger the genome, the slower is the rate of reassociation. Kohne (1970) explains this as follows:

FIGURE 25

RENATURATION SPECTRUM OF MOUSE DNA.

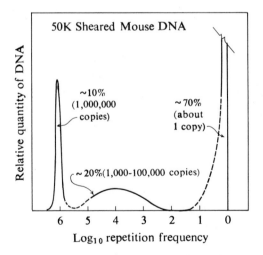

From Britten and Kohne (1968, Fig. 12). Copyright 1968 by the American Association for the Advancement of Science.

Thus, the DNA of a small virus SV-40 cut into ten pieces yields ten *different* pieces. The DNA of *E. coli* cut to the same size piece yields about 10,000 *different* pieces. If these preparations of DNA are adjusted to the same total DNA concentration and reacted under the reassociating conditions, the *E. coli* DNA takes about 1,000 times as long as SV-40 DNA for one-half the molecules to reassociate.

This is another way of saying that shearing a larger genome and adjusting to some standard concentration essentially reduces the "effective" concentration of reacting species. The factor of genome size is undoubtedly a dominant factor in determining the rate of this reaction, but it is not the only factor. However, Britten and Kohne have interpreted their rate data as if it were.

Of course, many factors influence the rate, but presumably all of them are held constant in the standardized conditions of the experiment. We shall discuss presently the fact that, when one is dealing with highly redundant sequences, the nature of the sequence itself has a significant influence on the rate of reaction and the stability of the reassociated complex. It is an additional factor that can influence the "effective" concentration of reacting species, and this factor cannot be held constant. It has not even been considered in the interpretation of these experiments.

Britten and Kohne interpret their rate data to mean that a fraction which reassociates, for example, 10^5 times faster than the main DNA contains sequences which have been repeated 10^5 times. From the known length of the genome and the number of copies so estimated, they calculate the length of the repeated sequences. Britten and Kohne (1968) state:

> It appears that, on the average, repeated sequences are not extremely short (not less than 200 nucleotides) and may be much longer than our fragments, which average perhaps 400 nucleotides.

After reading and musing about these experiments, I wrote the following passages, some of which appear in the review article by Jukes and Gatlin (1970).

Britten and Kohne believe that genes or segments of genes from the main DNA have been duplicated as many as a million times in events which they call "saltatory events." There is some evidence that at least some portion of the repeated DNA is transcribed into RNA (Britten and Kohne, 1970). How could the organism use so many copies of the same gene? Britten and Kohne (1968) express their feelings on this matter in the following statement:

> A concept that is repugnant to us is that about half of the DNA of higher organisms is trivial or permanently inert (on an evolutionary time scale).

This DNA may not be "trivial" or "inert" at all from an evolutionary viewpoint. It could represent a mechanism in higher organisms not only for reducing error but also for completely eliminating it at certain critical points in development and function. Suppose it is essential that there be no error in a particular tRNA molecule. If there are in the DNA a large number of identical templates for this tRNA which are transcribed, and if only a "perfect" molecule is functional, then it does not matter if a small percentage of the copies are defective. They could be eliminated from the system or simply not used, and perfect function would be statistically ensured at a critical point.

Insurance is not wasteful or trivial, just costly. Even so, it still seems that a million copies is extremely expensive insurance. However, it may be that the evolution of efficient controls lags behind the evolution of the duplication mechanism itself. A more efficient use of duplicate templates might involve a search procedure followed by a decision procedure which would select only perfect templates for transcription. With such a programmed mechanism of quality control, the need for a very large number of duplicate templates would decline in direct proportion to the efficiency of the selection procedure. Such mechanisms may be presently evolving.

On the other hand, we should consider the possibility that at least some of this rapidly reassociating DNA, particularly the most rapidly reassociating fractions, did not arise by exact duplication of 200 or more nucleotides of the main DNA, but rather as an evolutionary descendant of some form of naturally occurring, highly redundant polymer.

Such polymers were observed as early as 1960 when Sueoka (1960) discovered a satellite DNA in the crab, *Cancer borealis*, which is mainly an alternating ATATAT... sequence. However, there are significant quantities of the AA and TT doublets as well as 3% (C + G), so that the DNA is not a perfect alternating sequence (Swartz *et al.*, 1962). The redundancy of this DNA is 0.83 (see Table 3), which is "highly redundant" relative to the main DNA of all other living organisms. Cheng and Sueoka (1964) found that this highly redundant DNA comprised about 30% of the total DNA of widely differing tissues of *Cancer borealis*.

There are other observations of naturally occurring highly redundant DNA (or RNA). Bellamy and Joklik (1967) described an unusual RNA associated with reovirus. Although the main portion of the virus is double-stranded RNA, they observed a single-stranded RNA comprising about 15 to 20% of the total RNA which had a base composition of 87.8% A, 10.5% U, 1.4% C, and 0.3% G. This must be highly redundant, because such a base composition constitutes a sufficient (although not a necessary) condition for high redundancy. We would have a high R of the high $RD1$ type. Although there are not many such observations, where one can definitely say that the DNA or RNA has a high R value, there are a large number of observations of "satellite DNA" in a wide range of organisms including human DNA (Corneo *et al.*, 1970). Many of these satellite DNA fractions could well turn out to be highly redundant DNA if and when the nearest neighbor measurements are made for them or when they are sequenced.

The questions before us are: "What are the redundancy values of the rapidly reassociating fractions of Britten and Kohne?" "Is there a relationship between R and $c_0\tau$?"

Britten and Kohne (1968) show the monotone, poly U-poly A, in their Fig. 2, which is our Fig. 24, and note that its rate of reassociation is instantaneous under their experimental conditions. The redundancy of a monotone is 1, that of the main DNA is $\leqslant.11$. There is a range of about 10^{10} in $c_0\tau$ between these two extremes. $c_0\tau$ for mouse satellite is only about 10^3 times greater than the monotone. One cannot help wondering if this DNA has a redundancy value in the intermediate range between the monotone and the main DNA. Such a value could characterize a sequence varied enough to prevent self-reassociation through looping back, but redundant enough to associate with a large proportion of the reaction fragments, particularly if the percentage of bases reassociated is only 50% in some cases, as reported.

If the nearest neighbor frequencies of these fractions were to be measured, we could answer all these questions. If we find that the redundancy of the rapidly associating fractions is in or near the range observed for the main DNA of all living organisms, 0 to .10, then concepts such as "gene duplication" and "saltatory events" used by Britten and Kohne are relevant. However, if we find that this DNA is highly redundant, then these concepts may have to be substantially modified or even discarded as not meaningful for this type of DNA. We may have to consider the possibility that this DNA does not represent a million copies of a gene, but is instead a modified, highly redundant, naturally occurring polymer wherein shorter sequences are repeated, but not exactly, and the length of repetition may be difficult if not impossible to define. Simply stating the redundancy value may be a much more meaningful description.

For satellite DNA, whenever the two strands can be separated and the nearest neighbor frequencies determined for each strand, we should be able to sequence such DNA quite easily from the relative magnitudes of the doublet frequencies.

I then attempted to estimate just how short these repeated sequences might be. This involves a fundamental problem which we shall now discuss.

MAXIMUM REDUNDANCY

Let us analyze the properties of highly redundant sequences. First, let us list the various ways in which we can achieve a sequence whose redundancy is 1. This is the most highly ordered sequence under the entropy criterion since H_M must be 0. It may help to refer to the entropy scale, Fig. 5, throughout the following discussion. I also list below certain equations we will use again.

$$R \log a = D_1 + D_2 \tag{52}$$

$$D_1 = \log a - H_1 \tag{19}$$

$$D_2 = H_1 - H_M \tag{45}$$

$$H_M = -\sum_i \sum_j p_i \, p_{ij} \log p_{ij} \tag{43}$$

We may list three general cases.

Case A. $R = 1$ may be achieved entirely through D_1. Then $D_1 = \log a$, $D_2 = 0$, $H_1 = 0$, $H_M = 0$.

Case B. $R = 1$ may be achieved entirely through D_2. Then $D_2 = \log a$, $D_1 = 0$, $H_1 = \log a$, $H_M = 0$.

Case C. $R = 1$ may be achieved through a combination of D_1 and D_2. Then $D_1 + D_2 = \log a$, $H_M = 0$, but H_1 may have an intermediate value between 0 and $\log a$.

What are the properties of the sequences representative of each case? Case A is the monotone, since there has been maximum divergence from equiprobability of the letters. Case B is any permutation of the entire alphabet repeated indefinitely. For DNA, Case B is the sequence ATCGATCGATCG . . . or any permutation of it. Since the letters are all equally probable, $H_1 = \log a$ and $D_1 = 0$. Since $p_{AT} = p_{TC} = p_{CG} = p_{GA} = 1$ and all the other p_{ij} are zero, the expression for H_M vanishes and $D_2 = H_1 = \log a$.

For H_M to vanish, all the p_{ij} must be either 1 or 0 as in Case B. In Case C one way to accomplish this and still have a contribution to R from D_1 is to delete one letter from the alphabet and construct the sequence as in Case B. For example, for DNA, one such sequence is ATCATCATC . . . . This is a limiting case and, as R approaches 1, we approach such a sequence. An occasional G should appear in this long sequence to define the alphabet as $a = 4$. We shall discuss this matter in more detail in the last section of this chapter.

The analysis above makes possible a definitive description of satellite DNA. If the DNA is highly redundant, and if $RD1$ is high, i.e., the high R value is achieved with a large contribution from D_1, then the satellite will appear as a separate species in a density gradient centrifugation because its base composition will deviate significantly from the main DNA. However, if the highly redundant DNA is of the high $RD2$ type, then it will appear as a renaturation "satellite" as in Fig. 25. In both cases the DNA is highly redundant; they merely differ in their D-indices. The two criteria may overlap.

R-REDUNDANCY AND *S*-REDUNDANCY

Let us examine the two following sequences from an alphabet of two letters.

I: ATATATATAT...

II: AATTAATTAA...

Which sequence is the more highly ordered? Our "common sense" intuition tends to tell us that both sequences are highly nonrandom, but a simple calculation such as we have illustrated above quickly shows that for, sequence I, $R = 1$, whereas, for sequence II, $R = 0$. Thus, under the entropy criterion, sequence II is perfectly "random." *It is apparent that we are observing here a kind of ordering which is not characterized by the entropy measure. This ordering is due to a basic repeating unit in the sequence.* We can characterize such ordering very simply as the repeat length, L. The smaller L is, the greater the "ordering" under this criterion. For sequence I, $L = 2$ and, for sequence II, $L = 4$. Thus sequence I is more highly ordered under both criteria.

This is really the same situation we discussed in Chapter 3 as two kinds of redundancy, *R*-redundancy (simple repetition) and *S*-redundancy (Shannon's redundancy). The smaller the repeat length, the greater is the *R*-redundancy. There is obviously an interrelationship between *R*-redundancy and *S*-redundancy which we must now explore.

Although the two examples of naturally occurring, highly redundant DNA (or RNA) which we cited are characterized by extreme divergence of the base composition from equiprobability, this is not a necessary condition for high redundancy. For example, in Case B the sequence ATCGATCGATCG... has $R = 1$, although the bases are equiprobable. In fact, if any very short sequence is repeated indefinitely, the resulting polymer is highly redundant. Let us substantiate this statement.

Let us define $R \geqslant .5$ as "highly redundant." Since all naturally occurring DNA has $R \leqslant .11$, we may regard the evolutionary process (or source which originally assembled this DNA) as a random number or letter generator. The evolution of higher forms then proceeded as a divergence from these primordial random sequences. A perfectly random sequence under the entropy criterion is one with $R = 0$.

A perfect random number generator is often defined as one which emits sequences such that the n-tuples are uniformly distributed in n-space. For $n = 2$ and $a = 4$ this would mean that the 16 doublet sequences are all equiprobable. Hence we may regard D_2 as the divergence from equiprobability of the doublet sequences when $D_1 = 0$. As an example, consider again the simple case where $a = 2$. For the sequence ATATAT . . . , $R = 1$ whereas, for the sequence AATTAATT . . . , $R = 0$. In both cases $D_1 = 0$. In the latter case all four possible doublet sequences are of equal frequency whereas, for the alternating ATAT . . . sequence, the doublets AA and TT are completely absent. When we consider a long sequence composed of small repeating units, for an alphabet of 4 letters, it becomes clear that, if L is the length of the repeated sequence, L must be of sufficient length before the 16 doublets can all be represented equally. In fact, the minimum value of L for $a = 4$ and $R = 0$ is 16. If sequences of length shorter than this are repeated, R becomes greater as the sequence becomes shorter because not all of the doublet sequences can be equally represented.

To make this statement more quantitative I generated sequences of length $L = 3$ to $L = 100$ for $a = 4$, using a random number generator (the Berkeley CDC 6400 system generator). For each value of L, I generated a random sequence 100 times and calculated the mean and standard deviation of R. Figure 26 is a plot of the average R value versus L. Since L is a measure of the amount of repetition in the long sequence, the figure shows a functional relation between the S-redundancy and the R-redundancy of the sequence. Since smaller L means greater R-redundancy, Fig. 26 states that higher S-redundancy is associated with higher R-redundancy when the source is a random generator. The exact nature of the function must depend on the generator. There are a number of reasons why the generator used is not perfectly random, but it is sufficiently sophisticated for present purposes. With the random number generator used, the length of sequence repeated must approach 100 nucleotides before the R value approaches .05, the mean of the range of living organisms. It is quite clear in the case of the crab satellite DNA that its high S-redundancy is the result of a high R-redundancy.

All the statements above are true for a random generator or source. If the source is nonrandom, it is possible for R to approach 0 for much shorter L, as shown for example by the dashed line in Fig. 26. As we

FIGURE 26

AVERAGE *R* VERSUS LENGTH OF REPEATED SEQUENCE.

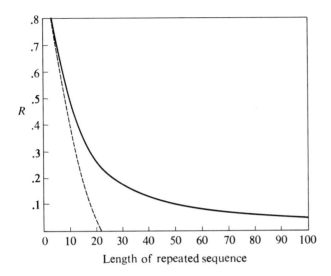

Length of repeated sequence

noted, if the sequence of symbols is chosen properly, the minimum length for $R = 0$ is $L = 16$.

This is an interesting situation. The "nonrandom" generator produces for short L a more random sequence under the entropy criterion. Our example sequence, AATTAATT . . . , also illustrates this situation. A "random" generator would not often produce a sequence with $R = 0$ for $L = 4$. The sequence of symbols must be rigidly controlled to produce this. We shall discuss this matter further in the next section, where we discuss the two measures of "randomness" being used here.

All of the preceding discussion is relative to a source with a memory of 1. It is true that, relative to a source with a memory of 3, the redundancy of sequence II is 1. In fact, for any source with a finite repeat length L, where the sequence is repeated exactly, $R = 1$ for a memory of $L - 1$. However, we encounter sources where the sequence statistics available do not allow consideration of a higher memory source. This is the existing situation for DNA.

We set out to estimate how short the repeat length might be in the rapidly reassociating fractions of Britten and Kohne. From Fig. 26 we drew the following conclusion (Jukes and Gatlin, 1970):

Hence, if we find that the rapidly reassociating DNA is highly redundant relative to the main DNA, this would mean that sequences of length shorter than 100, and possibly very short, have been repeated.

While this paper was in press, Southern (1970) reported that he had sequenced partially the satellite of guinea pig DNA and found it to be a polymer with a repeat length of 6. The basic sequence is CCCTAA. Preliminary investigations also indicate that the light satellite of the mouse has a repeat length of between 8 and 13 bases which includes the sequence TTTTTC. These sequences have been modified by point mutation so that they are not perfect repeating sequences. The situation is completely analogous to that of the crab satellite.

As we know, a good theory must not only explain a wide variety of empirical fact, but it must also be able to predict. This is our first fulfilled prediction.

NATURE OF THE SEQUENCE

I shall now show that, when one is dealing with highly redundant DNA, the nature of the sequence itself, i.e., the particular sequence of bases it contains, is an important factor in determining rate and stability of the reassociated complex. This factor has not even entered into the interpretation of these experiments as yet. Some workers (Southern, 1971; Sutton and McCallum, 1971) are already aware of this omission. They have shown that the fraction of mismatched base pairs in the final duplex, which is, of course, a direct function of the base sequence itself, has a strong influence on the rate of reassociation. I show below a related but additional effect of the base sequence on the reaction rate.

Classical work had shown that, when some of the base pairs in the DNA double helix are mismatched, i.e., are not the standard Watson-Crick pairs, about 1.5% base pair mismatches will lower the thermal stability of the complex by $1°C$. Britten and Kohne observed a strong reduction in thermal stability and a broadening of the range of dissociation of the *rapidly* reassociated complex. From hyperchromicity measurements they estimate that in some cases only half of the bases are paired.

They next introduce the concept of a "family" of sequences which are produced at a distinct time in the evolutionary history of the organism (the "saltatory event"). A family is composed of member sequences which are similar to but not identical with each other, each of which can reassociate with any other member of that family. These families of sequences show different degrees of similarity among the members, and the degree of similarity is judged solely on the basis of the thermal stability of the reassociated complex.

Kohne (1970) is careful to emphasize that the definition of a family, indeed even the definition of repeated DNA, is relative to the "criterion" of the experiment, i.e., salt concentration, temperature of incubation, etc. Under a stringent criterion, usually high temperature, only those families whose members are very similar can react and the complexes formed are thermally stable. When the reaction criterion is lowered, in the DNA of every higher organism examined thus far the result has been that a larger fraction of the DNA behaves as repeated DNA, i.e., reaction occurs but the complexes are poorly matched and have low thermal stability.

The explanation is offered that such families originated long ago and have "diverged" through the classical evolutionary mechanism. I am skeptical of the accuracy and completeness of such a classical interpretation. I illustrate my reasons with some simple example sequences.

Below are listed four arbitrary polymers with $L = 6$. Sequence I is, in fact, the one found for the guinea pig satellite:

I: CCCTAA/CCCTAA/...

II: TTTTTC/TTTTTC/...

III: TCGTCA/TCGTCA/...

IV: ATCGTA/ATCGTA/...

Let us write the Watson-Crick complementary sequence for each of these polymers on a separate strip of paper. When the complementary sequence is out of register by one base, we will call this the first register shift; by two bases, the second register shift, etc. The second register shift for sequence I is

CCCTAA/CCCTAA/...

GGGA/TTGGGA/TT...

TABLE 10

PERCENT MISMATCH VERSUS REGISTER SHIFT FOR EXAMPLE
SEQUENCES

Sequence	Register Shift	% Mismatch
I	1	50
	2	83 1/3
CCCTAA . . .	3	100
	4	83 1/3
	5	50
	6	0
II	1	33 1/3
	2	33 1/3
TTTTTC . . .	3	33 1/3
	4	33 1/3
	5	33 1/3
	6	0
III	1	100
	2	100
TCGTCA . . .	3	33 1/3
	4	100
	5	100
	6	0
IV	1	83 1/3
	2	100
ATCGTA . . .	3	66 1/3
	4	100
	5	83 1/3
	6	0

We note that for the unit repeat length of 6 there are five out of six base pairs which are not the standard Watson-Crick partners, i.e., they are mismatched. Table 10 lists the percentage of mismatches for each register shift of each of our four example sequences.

Whether or not these sequences reassociate depends on the "criterion," i.e., the entire set of reaction conditions. Let us take the criterion that a mismatching of $33\frac{1}{3}\%$ is permissible and a stable complex forms under this criterion. Since some complexes are only 50% matched, this is a conservative criterion.

If we assume that there is an equal probability of collision of the strands in any register, under the $33\frac{1}{3}\%$ criterion sequence II has a 100% probability of reassociation, sequence III a $33\frac{1}{3}\%$ chance,

TABLE 11

RATE OF REASSOCIATION AND STABILITY OF COMPLEXES FOR
EXAMPLE SEQUENCES UNDER 33 1/3% CRITERION

Sequence	Probability of Reassociation	% Mismatch	Fraction of Successful Collisions
I	1/6	0	1
II	1	0	1/6
		33 1/3	5/6
III	1/3	0	1/2
		33 1/3	1/2
IV	1/6	0	1

and the other two sequences only a $16\frac{2}{3}\%$ chance of reassociation. This is a measure of their relative rates of reassociation. Furthermore, there is a significant difference in the stabilities of the reassociated complexes. For example, for sequence II, $\frac{5}{6}$ of the collisions will result in complexes which are $33\frac{1}{3}\%$ mismatched, and $\frac{1}{6}$ of the reassociated complexes will be in perfect register. However, for sequences I and IV only those collisions which result in perfect register will produce stable complexes. Table 11 summarizes this situation for all sequences.

We could extend these calculations and repeat the entire procedure for a different stability criterion, different repeat lengths, etc. However, the simple example given above is sufficient to show that the rate of reassociation and the stability of the reassociated complex are a function of the nature of the sequence itself. All of these sequences have the same repeat length and all are composed of identical unit sequences. There is no difference in "similarity" among the members of a "family." No "divergence" of any kind has taken place. Yet they would reassociate at different rates and form complexes with different thermal stability.

Britten and Kohne completely neglect consideration of this factor in the interpretation of their experimental data. Both of the factors they consider, the repeat length and the evolutionary divergence, are undoubtedly important factors in determining rate and stability, perhaps dominant factors in most cases, but the nature of the sequence is an *additional* factor which should be considered. It could make a significant difference in the interpretation of these experiments, particularly for the highly redundant DNA.

The "family" concept of Britten and Kohne is really an attempt to define a set of sequences which belong to a particular source. We discussed criteria for selecting such a set of sequences in the section on ergodic Markov sources in Chapter 3. The source concept has not been applied in classical evolutionary thought, but it is an extremely useful conceptual device.

There are various ways to describe the degree of similarity of sequences, and various measures have been defined and applied extensively to protein sequences with the primary objective of determining which sequences descended from the same ancestral sequence. When the degree of similarity between two sequences is quite high under any of these measures, the conclusion is compelling that they are evolutionary descendants of the same sequence which has been modified slightly by point mutation, insertion, or deletion. However, when this degree of similarity begins to fade, a grey area appears where it is difficult if not impossible to decide if the sequences have descended from the same ancestral sequence (homology) or if they have arisen independently (analogy).

It can be shown that random sources under rather light constraints can generate sequences with weak degrees of similarity. It may become more useful in evolutionary thought simply to regard sequences with defined degrees of similarity as the output of the same source without emphasis on the exact *physical* nature of this source. Of course, all the sequences which have descended recently from the same ancestral sequence could be regarded as belonging to the same source. The "families" of rapidly reassociating DNA are the outputs of individual sources. Since it has been observed that closely related species do not share the same families, these families have been generated by sources in a relatively short time on an evolutionary time scale and different individuals possess different sources. As I have shown, we should not regard the thermal stability of the reassociated complex *alone* as an accurate measure of the degree of similarity of sequences from these sources.

R-REDUNDANT SOURCES

Our investigation of these beautiful experiments leads us into a fundamental extension of our notions of the "randomness" of sequences.

The random number "generator" which I used to investigate the repeat length of R-redundant sequences is actually a source, an entity which emits a sequence of symbols.

Up to the present chapter, we have been on firm classical ground in regard to our basic measure of the "ordering" or "structuring" of a sequence of symbols emitted by a source. Shannon's fundamental functional form, $-\Sigma_i p_i \log p_i$, the entropy, and all of the quantities which we have derived as a consequence of it, are the statistical averages of $-\log p_i$ which is a random variable on S_1. The concept of a statistical average requires a sufficiently long sequence to be meaningful. The random source of Fig. 26 does approach $R = 0$ as L becomes larger.

However, the observation of sequences with a short repeat length in the DNA of living organisms confronts us with a new and deep fundamental problem. There is nothing to prevent us from calculating R and D for these R-redundant sources because they do emit long sequences; but we are immediately touching upon new and unexplored concepts.

All the facts (the p_i and p_{ij}) necessary to compute R and D are contained within one unit segment of the sequence, and this segment may be quite short. (One must count the doublet consisting of the last symbol in the sequence followed by the first symbol as the "connecting doublet," but this "information" still resides within one unit segment.)

This situation is reminiscent of the "signature" concept of Henry Quastler (1964) wherein a short sequence within a longer sequence contains all the "vital" information. We can also describe this situation in terms of the concepts of stored information and potential information which we developed at the outset of this book. The situation is completely analogous to the illustration from Cherry (1957) of the book binder's error which bound all of page one in one book, all of page two in the next book, etc. After reading the first page of any book, no further knowledge or meaning can be transmitted because the potential information has dropped to zero. As I noted in the section on language in Chapter 3, if we carry either variable to the extreme, we lose the meaning. However, if the only meaning to be conveyed is contained in the basic repeating unit, this extreme R-redundancy will assure statistically 100% fidelity of the message. This is why I believe that such sequences serve vital control functions in the living system.

Life is versatile. It uses every kind of redundancy in the most advantageous way.

We are faced with the problem of interpreting the concept of randomness for such sequences and their sources. We touched upon this problem earlier in this chapter when we noted that for short L a "nonrandom" source is required to produce a "random" sequence (one with $R = 0$). This apparent inconsistency is due to the fact that we are now dealing with two measures of the ordering or degree of randomness of sequences (or the sources which emit them). One measure is the classical entropy criterion. A random source is one with $R = 0$. The second measure is the repeat length, L. If $L = \infty$, i.e., the source never repeats itself, it is, in a certain sense, less highly ordered than when L is finite. If $L \neq \infty$, let us call the source repetitive (or R-redundant) and, if $L = \infty$, nonrepetitive (or non-R-redundant).

If we consider the criteria that R may be either zero or nonzero and L may be either infinite or finite, there are four possible types of sources. Even though the difference between them is quantitative, not qualitative, it is nevertheless instructive to examine these four sequence types. They are listed in Table 12.

If $R \neq 0$ and $L \neq \infty$, we have a nonrandom repetitive source, as exemplified by the "highly redundant" DNA of the guinea pig or mouse satellite. If $R = 0$ and $L = \infty$, we have a random nonrepetitive source. Computer system random number generators attempt to approach the behavior of such sources. If $R \neq 0$ and $L = \infty$, we have a nonrandom nonrepetitive source whose sequences may be regarded as sequences originally from a random nonrepetitive source that have

TABLE 12

BASIC SOURCE TYPES		
Ordering Criterion	*Descriptive Words*	*Example Sequence or Source*
$R \neq 0, L \neq \infty$	Nonrandom repetitive[a]	Satellite DNA of Guinea Pig
$R = 0, L = \infty$	Random nonrepetitive[b]	Random number generator
$R \neq 0, L = \infty$	Nonrandom nonrepetitive	Main DNA
$R = 0, L \neq \infty$	Random repetitive	AATTAATT ...

[a] The word repetitive may be replaced with the word R-redundant.
[b] Nonrepetitive = non-R-redundant.

diverged from $R = 0$. At this time the main DNA of living organisms *appears to be* of this type, although it is too early to say that there are no basic repeating sequences in this DNA. Finally, for $R = 0$ and $L \neq \infty$ we have a random repetitive source whose output would consist of repeated sequence of finite length with $R = 0$. The sequence AATTAATT . . . is of this type.

The constraints on the source may arise from either R or L. In the case of our AATTAATT . . . sequence, it is clear that the constraints which must be placed upon the source to produce this sequence arise from its repetitive nature, its R-redundancy, which is a very real type of "nonrandomness." As L becomes smaller, these constraints become more stringent.

It is apparent that the discussion above extends our classical notions about the "randomness" of sequences of symbols. It also presents the possibility that, since the different types of DNA are the output of different types of sources, the physical evolutionary hardware corresponding to the different sources may be quite different. For example, the main DNA appears to be the output of an $R \neq 0$, $L = \infty$ source, and the satellite DNA, the output of an $R \neq 0$, $L \neq \infty$ source. Hence their evolutionary origins, i.e., the physical hardware of these sources, could be quite different.

I will continue to refer to the satellite DNA with short L from the $R \neq 0$, $L \neq \infty$ source simply as "highly redundant" DNA because it has both high S-redundancy and high R-redundancy. There is undoubtedly a source within the living cell that produces sequences of copies of genes which code for protein. This is an $R \neq 0$, $L \neq \infty$ source also, but the repeat length would be of the order of magnitude of a hundred or more and the S-redundancy would in general be lower. This DNA is of intermediate redundancy. It should be found in the DNA fractions of intermediate reassociation rate.

FUNCTION OF HIGHLY REDUNDANT DNA

The informational capacity of highly redundant DNA is unique and quite distinct from that of the main DNA. Because of its high R-redundancy and high S-redundancy, it is a highly accurate genetic message. But this extreme fidelity must have been purchased at the expense of significant loss of message variety. Therefore such sequences may be regarded in a very general way as "simple but accurate" genetic messages.

Now, if one wishes to give control commands to a computer such as STOP, START, PAUSE, GO TO, he does not require a large vocabulary; but it is imperative, particularly if the function of a system depends on them, that such instructions be transmitted without error. To me the conclusion seems compelling that highly redundant DNA sequences are part of the language from the master program which controls lower functions such as protein synthesis, i.e., these sequences are *control* sequences.

In 1968 (Gatlin, 1968) I stated:

> I previously speculated (Gatlin, 1966) that highly redundant DNA sequences might serve instructional purposes because of their unique informational character.
>
> The significance of redundant sequences is their capacity to combat error. A protein may survive several amino acid changes, but because [the meaning of] instructional sequences and molecules would have a much lower error tolerance it makes sense to use redundant sequences for instructional purposes. Perhaps the quantitative values of R and D play a role in this function.

As I was writing the final revision of this book, an article by Sivolap and Bonner (1971) appeared. They separate the DNA into three fractions, the "most repetitive" DNA, the "middle repetitive" DNA, and the "unique" DNA. The chromosomal RNA hybridizes most extensively with the most repetitive DNA, significantly less with the middle repetitive DNA, and almost not at all with the unique DNA. Sivolap and Bonner state:

> We show below that chromosomal RNA interacts with the repetitive sequences. Since chromosomal RNA is established as a control element of the chromosomes of higher organisms, it follows that the repetitive sequences of the DNA are likewise, in part at least, also control elements.

This supports my 1966 and 1968 speculations. This work also supports the theory of Britten and Davidson (1969) which likewise postulates

FIGURE 27

MAX MIN OF DNA CONTENT

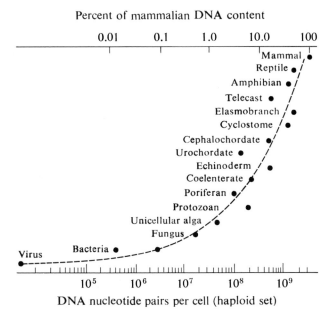

Percent of mammalian DNA content

DNA nucleotide pairs per cell (haploid set)

From Britten and Davidson (1969, Fig. 3). Copyright 1969 by the American Association for the Advancement of Science.

a control function for the repeated DNA. Their model, like the model of Jacob and Monod (1961), postulates specific hardware and circuitry of the control system involving such components as producer genes, receptor genes, activator RNA, integrator genes, sensor genes, and batteries of genes. I will not review the hardware. The model requires that there be many copies of genes which produce mRNA. As we have discussed, this DNA would be of the $R \neq 0$, $L \neq \infty$ fundamental type but generally not so high in R-redundancy or S-redundancy as the satellites.

It is known that very closely related species may not share the same families of repeated DNA (Kohne, 1970). Therefore these families are the output of evolutionary sources which periodically, and in a relatively short evolutionary time period, produce a new class of sequences. Figure 27, which is reproduced from Britten and Davidson (1969), shows that as higher organisms have evolved they have maximized the minimum DNA content of their cells. Britten and Davidson grasp the concept that this additional DNA is being used in higher and higher levels of control in the programming functions. They simply express this concept in different language.

Another requirement of the model is that repeated sequences be scattered throughout the genome rather than localized in one segment of it. This is found experimentally (cf. Britten and Davidson, 1969). In our language, it is obvious that the control instructions must be located sequentially throughout the program. Mouse satellite DNA appears to be localized near the centromeres of the cell (Jones, 1970). Perhaps many programming operations occur in these complex and still fundamentally mysterious cellular elements.

The significance of the appearance of new families periodically is that there must be a master program which can generate subprograms, write them itself, and even generate the hardware (the control sequences) necessary to implement them. Britten and Davidson express the same concept in their language:

> Both the rate and the direction of evolution (for example, toward greater or lesser complexity) may be subject to control by natural selection.

The words "natural selection" play a role in the vocabulary of the evolutionary biologist similar to the word "God" in ordinary language. If we do not know the nature of first causes, we rely upon words such as these. My words, "master program," serve a similar function, but I feel that they are more useful in leading us into an understanding of the hierarchy of causes.

There are two possibilities regarding the mechanics of function of control sequences. A specific sequence may be required or some variation in the sequence may be permissible. If a particular sequence of bases is required, high R-redundancy will ensure fidelity. This is the principle behind gene duplication, as discussed early in this chapter. If variation is permissible, not only could mutations in the sequence be tolerated, but also these changes would be reversible and subject to control. Britten and Davidson postulate such a nature for their receptor genes. Both types of sequences probably exist and serve different control functions.

TOWARD A THEORY OF MEANING

In describing the informational properties of short sequences in this chapter, we have departed from classical concepts. In reality the highly redundant DNA is a type of *highly ordered DNA*, and I believe that other measures of the ordering of short sequences based upon their symmetry properties will be found to be useful in describing control sequences.

As the signature principle of Quastler suggests, it is perfectly natural to attach informational measures to very short sequences. In fact, Quastler (1964) defines such a measure as

$$\sum_i n_i \log p_i \qquad (76)$$

where the n_i are actual character counts in the short sequence and the p_i are theoretically estimated ideal probabilities.

Let us go a step further. I see no reason to preclude the possibility that a single symbol can alter significantly, even drastically in some cases, the informational capacity of the sequence in which it is imbedded. A simple example can show that the insertion of single symbols in a long sequence can alter the entropy information measures because they, in essence, redefine the alphabet. Let us again use our binary example sequences. If we observe very long sequences such as ATATAT... or AATTAATT..., in the absence of any further information about the source we would assume that these sequences are the output of a source with an alphabet of two letters, i.e., $a = 2$. We would calculate their informational parameters and obtain the values described as "before" in Table 13.

TABLE 13

REDEFINITION OF THE ALPHABET							
Sequence	H_1	H_M	D_1	D_2	R	$RD1$	
I							
ATATAT...	1	0	0	1	1	0	before
	1	0	1	1	1	.5	after
II							
AATTAATT...	1	1	0	0	0	0	before
	1	1	1	0	.5	1	after

However, if we observe that upon rare occasions the letter C or G makes its appearance in the output of the source, we are faced with an interesting situation. We are now forced to conclude that $a = 4$ for the source. When we adjust our calculations accordingly (now $\log_2 a = 2$ whereas before, $\log_2 a = 1$), we obtain the set of values listed in Table 13 as "after." The redundancy of sequence II has changed from 0 to .5. We note that H_1, H_M, and D_2 are invariant, because the p_{ij} and the p_i are not significantly changed by the rare C or G. D_1, R, and $RD1$ (or $RD2$) are variable. However, if as in sequence I, $H_M = 0$, this fixes R at unity since $R = 1 - (H_M/\log a)$.

This example shows that information measures, like probabilities, are relative to one's previous state of knowledge. This is the reason for defining conditional probabilities. For similar reasons, we should regard all information measures as conditional information relative to one's definition of the alphabet or source. In the case of DNA or protein sequences, there is no ambiguity problem regarding the definition of the alphabet.

Our example illustrates the tremendous influence the appearance of a new letter in the alphabet can have on the informational capabilities of a sequence. I believe that this is precisely why we observe all the modified bases in tRNA molecules. They have enlarged the alphabet; this endows them with unique, additional informational capacity enabling them to serve more sophisticated informational roles. They are the nucleic acid molecules which go *On Beyond Zebra* (Seuss, 1955).

We are really here taking the first halting steps toward a theory of meaning. We defined information at the outset as the capacity of a sequence of symbols to convey meaning, not the meaning itself. There may be reliable ways to define measures of the capacity of short sequences to convey meaning, a property which classical theory attributes only to long sequences. Different kinds of meaning require different kinds of capacities and hence different informational measures.

8

A disagreement may be the shortest cut between two minds. —Kahlil Gibran, Sand and Foam

THE INFORMATION DENSITY OF PROTEIN

THE PROBABILITY PROBLEM

Up to this point we have considered only the information density of DNA. We shall now consider the information density of protein.

A fundamental problem immediately presents itself. There are only four bases in DNA, and DNA molecules are very long, from about 10^4 to 10^9 bases. Thus we are reasonably secure in regarding the 16 doublet frequencies measured by the nearest neighbor experiment as probabilities. On the other hand, there are 20 different amino acids in protein, hence 400 doublet sequences; and all proteins are quite short compared to nucleic acids, of the order of magnitude of a few hundred amino acids or less for most proteins. Furthermore, there are only a limited number of protein molecules. For example, cytochrome c molecules are similar enough in all eukaryote organisms that we may regard them as evolutionary descendants of one prototype molecule or as the output of a single source. Thus, although we have a great deal of amino acid sequence data today, if we were to count the 400 doublet frequencies for all these data, it would be doubtful whether these numbers were representative probabilities. They might reflect the relative proportion of experimental measurements of a particular type of protein molecule. If we try to select a random sample of molecular types, we seriously limit the length of the sequence data.

Before 1971 no one had ever calculated D_2 for proteins but, on the basis of the sequence data available in 1958, Yčas concluded that there was no obvious evidence for independence of amino acids in proteins. Some more recent work agrees (Krzywicki and Slonimski, 1966) and some disagrees (Jones and Dayhoff, 1972).

Reichert and Wong (1971) have attempted to calculate D_2 for protein, $D_2(P)$, using nonclassical information measures and the formula of Christensen (1968) to estimate these probabilities which we cannot count. More work will have to be done before we know the value of these estimates. From a study of cytochrome c sequences they conclude that $D_2(P)$ for vertebrates is greater than that for lower organisms. This is a most acceptable conclusion. Let us derive it.

THE RANDOM SOURCE

In the classical search for restrictions on amino acid pairs, tables of doublet frequencies were constructed from all the available protein sequence data. To date only 23 of the 400 possible doublets are missing from this table (see Yčas, 1969). After classical statistical analysis, the conclusion was drawn that there are no forbidden pairs and no evidence for a nonrandom distribution of nearest neighbors. What this classical work essentially says is that all protein sequences are the output of a completely random "master" source with $R = 0$ where all n-tuples have an equal probability of emission. Let us assume this conclusion is correct.

It does not follow therefore that protein sequences are "random" sequences with $R = 0$ and $D_2 = 0$. This is the classical error. In fact precisely the opposite conclusion follows. We proved this in Chapter 7 where we showed that a completely unconstrained generator produces short sequences with high redundancy. The length of protein sequences is short relative to the size of the alphabet. Hence if we wish to consider a first-order Markov effect, there are 400 possible doublet sequences and most ordinary proteins are not even this long. To even attempt to calculate D_2 for ordinary protein sequences is roughly equivalent to attempting to calculate $D_2(DNA)$ from DNA sequences of length less than 16. It is clear that we must regard any values obtained with somewhat different perspectives than informational values for DNA.

Let us return to our random generator. This time we will generate sequences of lengths 10 to 340 from an alphabet of 20 letters. This

FIGURE 28

R VERSUS LENGTH OF SEQUENCE FOR PROTEIN RANDOM GENERATOR.

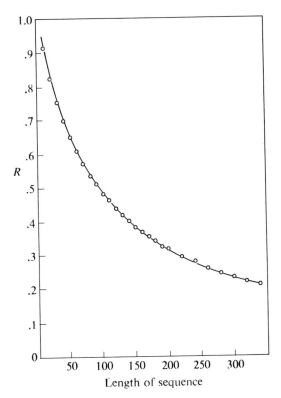

Length of sequence

covers the range of most common proteins. A plot of R versus L is shown in Fig. 28. The functional behavior is very similar to that for DNA. This function predicts that if proteins are the output of an unconstrained generator, for sequences of the order of magnitude in length as most common proteins, the shorter the length, the higher the redundancy of the sequence.

Figure 29 shows a plot of D_2 and D_1 versus length. D_1 drops rapidly and approaches zero asymptotically. The means of D_1 and D_2 stabilize at about length 50. At this length the standard deviation is about .1 and declines thereafter. For R, $\sigma \cong .01$ at length 50 and remains stable. Average R, D_1, and D_2 values were calculated from a sample of 100 randomly generated sequences for each length point.

FIGURE 29

D_1 AND D_2 VERSUS LENGTH OF SEQUENCE
FOR PROTEIN RANDOM GENERATOR.

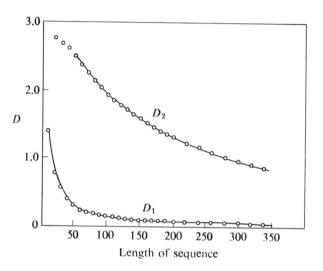

Without exception, the cytochrome c sequences of vertebrates are shorter than those of the lower organisms. For vertebrates the length is 103 to 104, for insects, 107, and for all other lower forms, 107 to 112. This simple analysis predicts that the higher organisms will have more highly ordered protein sequences under the entropy criterion, and the generator or source does not have to be *completely* random for this analysis to apply. In fact it could be under considerable constraint. The function would then follow a different path, depending on the nature of the constraint; but in the length range of ordinary proteins, the R and D_2 values must be monotonically decreasing functions of the length. It therefore follows that the ordering of cytochrome c sequences for either $R(P)$ or $D_2(P)$ is: vertebrates > insects > lower organisms and plants.

From Figs. 28 and 29 we can estimate that if the generator is unconstrained, the order of magnitude values for sequences of length about 100 is $R(P) \approx .5$ and $D_2(P) \approx 2$ bits. These are highly ordered sequences under the entropy criterion.

Figure 30 is a plot of R versus D_2 for our protein random generator. After $R \leqslant .4$ and $D_2 \leqslant 1.5$ approximately, which corresponds to a

FIGURE 30

R VERSUS D_2 FOR PROTEIN RANDOM GENERATOR.

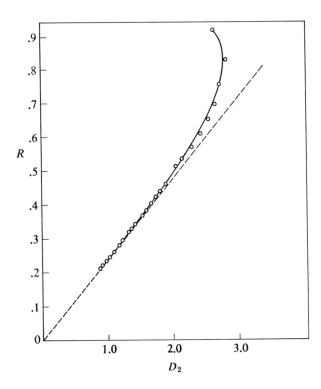

length of about 150, R versus D_2 becomes linear with $\alpha = .24$ and $\omega = 0$. The theoretically predicted slope for $D_1 = $ constant is $1/\log_2 20 = .23$. Thus the protein random generator also predicts the $RD2$ vector in suitable length ranges.

THE FIRST THEOREM

There is yet another way to predict from basic principles that the R and D_2 values of protein sequences must follow the same general ordering as the DNA values. The first theorem of information theory essentially states that the entropy cannot decrease in a simple mapping process, or equivalently, the information density cannot increase. If we assume sufficiently long sequences, for protein-coding DNA the following inequality holds:

$$I_d(\text{DNA}) \geqslant I_d(\text{P}) \tag{77}$$

From equation (56)

$$R(\text{DNA}) \log_2 4 \geqslant R(\text{P}) \frac{\log_2 20}{3} \tag{78}$$

(The factor of 3 is required since there are three DNA symbols per protein symbol.) We obtain the conversion inequality

$$R(\text{DNA}) \geqslant .72031 R(\text{P}) \tag{79}$$

Or if under certain circumstances we wish to consider the word STOP as a part of the protein language, $\log_2 20$ is replaced with $\log_2 21$ and we obtain

$$R(\text{DNA}) \geqslant .73205 R(\text{P}) \tag{80}$$

Therefore on a linear information density scale, the DNA space must lie above the protein space. In the mapping process from DNA to protein, I_d can either decrease or remain constant. The latter case may be regarded as a transformation occurring at the boundary of the information spaces. We have observed that for double-stranded DNA the I_d values of vertebrates are above those of lower organisms. Let us assume that this same relative ordering will be found for single-stranded, protein-coding DNA. When we map from the DNA to the protein space, if the $R(\text{P})$ values do not have this same relative ordering, the vertebrates would lose more information density in the mapping process than lower organisms. This would mean that lower organisms were more efficient information processing systems than higher organisms, an inherently unacceptable conclusion. We conclude again that the information density of protein sequences must be higher for vertebrates than for the lower organisms. Our consideration of the random source has given us an estimate of the magnitudes of these values relative to this source.

We must never forget that all information measures are relative to the source or information space we have defined, just as, in classical probability theory, all probabilities are relative to our definition of sample description space.

In 1969 there appeared a most significant paper by Temple F. Smith, the first work to make a transformation from the DNA to protein space and then calculate an information function on the protein space. The result is our first quantitative insight into the old and classical mystery as to why all vertebrates have about the same base composition of their DNA, roughly $42 \pm 2\%$ (C + G).

SMITH'S CALCULATION

S_1 is the basic sample description space which defines the alphabet. This is an extremely fundamental matter. As we showed in Chapter 7, the values of the entropy parameters depend vitally on the definition of the alphabet. S_1 may contain 4 elements which we would identify with the four DNA bases. Let us call this 4-space (4-S). If S_1 contains 20 elements, the 20 amino acids, this is a 20-space (20-S) and, if STOP is added as a word, we have a 21-space (21-S). Thus when $a = 4$ we have a DNA alphabet, hence a DNA language, hence a DNA space. When $a = 20$ or 21 we have a protein alphabet, language, or space. It is possible to calculate the information parameters on (4-S), (20-S), or (21-S), using either DNA or protein data. In the case of DNA we have the nearest neighbor data, and for proteins we have amino acid sequence data. Let us indicate the source of the data as a superscript. Let us refer to a, the number of letters in the alphabet, as the "size" of the space and reserve the word, order, to refer to the length of the n-tuple under consideration. There are many higher-order spaces, S_n, defined on an S_1 of a given size.

I first calculated (Gatlin, 1968) H_1, H_M, D_1, and D_2 on (4-S)^{DNA} (a DNA space from DNA nearest neighbor data).

In a very general sense, one can calculate informational parameters on a space of any alphabet size from either type of data, provided one knows the transformation rules connecting the spaces. In fact, one can calculate informational parameters on an information space using no data whatsoever but only certain theoretically defined informational functions, as we shall see presently.

Smith has calculated a number of quantities on a protein space with $a = 21$, i.e., on (21-S). Smith's original calculations utilized an incorrect codon dictionary with 22 words. G. B. Weiss (1970) has already criticized Smith for this. However, both Smith and I have recalculated, using the correct 21-word dictionary, and I confirm his values for the quantity $HR(\lambda)$ which we shall presently describe.

Smith makes the following basic assumptions in the calculation of $HR(\lambda)$.

1. The frequency of occurrence of a given amino acid in protein is directly proportional to the sum of the frequencies of its corresponding codons in DNA.

2. The genetic code correctly describes the transformation from (4-S) to (21-S).

3. Smith assumes $A = T$, $C = G$, thus making the single-base frequencies representative of double-stranded DNA.

4. $D_2 = 0$ for DNA; hence the codon frequencies can be calculated as the product of the single-base frequencies, i.e., the "random" frequencies.

The codon frequencies calculated under assumptions 3 and 4 are then summed under 1 and 2. For example, the frequency of phenylalanine is the sum of the frequencies of the codons UUU and UUC. From this distribution, Smith calculates H_1 on (21-S), which he calls $HR(\lambda)$. Since he assumes perfect Watson-Crick complementarity between the bases, this makes $HR(\lambda)$ a function of only one independent variable, λ. He then plots $HR(\lambda)$ versus λ as shown in Fig. 31. There is an absolute maximum in this function of $HR(\lambda) = 4.24$ bits at $\lambda = 43\%$ (C + G). I confirm this calculation.

This is a fascinating result because it is within the range where all vertebrate DNA is found. Unfortunately, Smith's paper contains some rather distracting inaccuracies. I have reviewed them in detail (Gatlin, 1971). However, the maximum in $HR(\lambda)$ is significant and correct and the inaccuracies in the paper cannot and should not detract from its significance.

Let us make sure that we understand precisely what has occurred in this calculation. No experimental data of any kind have been used. A transformation has been made from (4-S) to (21-S) where assumptions 1 and 2 are the transformation rules and assumptions 3 and 4 are additional assumptions about the nature of the original space, (4-S).

Jack Lester King (1971), after musing about this calculation, proposed an ingenious way to remove assumption 3. This is desirable

FIGURE 31

H_1 ON (21-S) VERSUS $\lambda = \%(C + G)$.

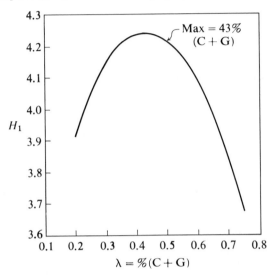

$$\lambda = \%(C + G)$$

because *in vivo* only one strand of DNA codes for protein and the single strand is not under the Watson-Crick constraints. Smith's calculation involves only one degree of freedom. King took three degrees of freedom or three independent variables for the base composition of DNA. They are representative of the single strand. Then, utilizing classical optimization techniques, he optimized these variables to maximize the entropy function H_1 after transformation from (4-S), using the same transformation rules as Smith, namely, assumptions 1 and 2. This calculation, of course, must involve assumption 4. The optimum values of the base composition which maximize H_1 are $A = 32.2$, $U = 23.3$, $C = 19.4$, and $G = 25.0$. The percentage of $(C + G)$ is 44.4%, which is higher than Smith's maximum. Also King left out the STOP codons *and* the amino acid arginine for biological reasons. Hence he calculated H_1 on (19-S).

It is also possible to remove assumption 4 by injecting the experimental nearest neighbor data into the calculation of the triplet frequencies. This, however, automatically reinjects assumption 3 since the nearest neighbor data are double-stranded data and closely approximate the Watson-Crick relations. When we do this, we have a value of

H_1 on $(21\text{-}S)^{\text{DNA}}$. Smith calls these values H^{Obs}, or H "observed." Even though the triplet frequencies are calculated using the DNA nearest neighbor data, hence the first-order Markov transition probabilities, once the amino acid probabilities are obtained under transformation rules 1 and 2 above, these probabilities are *a priori* probabilities (the single-letter probabilities) on $(21\text{-}S)^{\text{DNA}}$ and tell us nothing about the sequencing of amino acids in protein. Hence the quantity $H^{\text{Max}} - H^{\text{Obs}}$, which Smith calls the "information density," is a form of D_1 on $(21\text{-}S)^{\text{DNA}}$. Smith defines a maximum slightly below the upper theoretical bound of log a which I use to calculate D_1 on $(4\text{-}S)^{\text{DNA}}$.

All of Smith's H^{Obs} values are incorrect because he calculates the codon frequencies backwards (Gatlin, 1971). I have plotted a number of correct H^{Obs} values in Fig. 32. This figure should replace Smith's Fig. 1, which is incorrect. The H_1 values lie mainly on or slightly above the "random" curve of $HR(\lambda)$. Therefore Smith's result to which he attaches considerable significance, namely, that none of his H^{Obs} values lies above the random curve, is invalid. His calculations of H_{M}

FIGURE 32

H_1 ON $(21\text{-}S)$ VERSUS λ WITH EXPERIMENTAL
POINTS OF H_1 ON $(21\text{-}S)^{\text{DNA}}$.

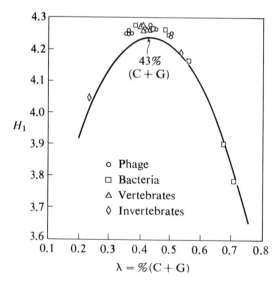

$(6, \lambda)$ are also all invalid because of the same error. This invalidates virtually all of Smith's evolutionary discussion, including his Fig. 2 and everything which follows it.

Again these inaccuracies in Smith's paper should not detract from the significance of the result of the maximum in H_1 on $(21\text{-}S)$. King's calculation confirms this result and improves on it. When we examine the assumptions underlying these calculations, even with 3 removed, the interesting question is why such a meaningful result is obtained at all.

There is no reason to expect a direct proportionality between codon frequencies and amino acid frequencies. We know that cells contain many copies of some proteins, while others are rare. Thus some DNA templates could be used repeatedly. Furthermore, only a certain percentage of the DNA is used for the synthesis of protein, and this percentage may turn out to be quite small, particularly in higher organisms. King and Jukes (1969) believe that "it is probable that not much more than 1% of mammalian DNA codes for proteins." Also the codon frequencies calculated under this model are overlapping frequencies whereas the genetic code is a nonoverlapping code. Yet under these imperfect assumptions and transformation rules Smith obtains his interesting result.

A possible answer to all these objections is that the informational principle with which we are dealing, namely the maximization of $H_1(P)$, was operative at a very primitive stage in the evolution of living systems, and all the assumptions we have made are reasonable approximations to the conditions existing at some time in this process when the genetic code was laid down, and with it, the optimum base composition.

Let us compare Smith's values for H_1 with those calculated from experimental amino acid composition data. Table 14 lists H_1 which I have calculated on $(20\text{-}S)^{\mathrm{DNA}}$ utilizing only assumptions 1, 2, and 3. This is essentially Smith's "H^{Obs}" value with the STOP codons removed from the dictionary. I wished to compare this value with H_1 calculated directly from the experimental data which, of course, do not contain a measurement of the frequency of the STOP codons. Considering all the assumptions involved in this calculation, I think the order-of-magnitude agreement is unexpectedly good.

Some of the data sets in Table 14 actually contain entries for only 18

TABLE 14[a]

COMPARISON OF H_1 THEORETICAL WITH EXPERIMENTAL VALUES

Organism	H_1 Theoretical	H_1 Experimental
T2	4.16	3.97
T4	4.15	3.96
λ	4.17	4.03
φ-X-174	4.17	4.10
B. subtilis	4.17	4.08
E. coli B_a	4.16	4.08
A. aerogenes	4.09	4.04
M. phlei	3.87	4.01
M. lysodeikticus	3.76	3.90
Chlamydomonas	4.13	4.00
Tetrahymena	3.93	4.06

[a] H_1 theoretical was calculated from a 20-word dictionary omitting the STOP codons for comparison with the 20 experimental amino acid frequencies. When aspartic acid and asparagine were analyzed together, I assigned half the value to each. Similarly for glutamic acid and glutamine. Data are from Fitch (1964) and Sueoka (1961).

or 19 amino acids. The rarer amino acids have been omitted from the analysis. However, the paper by Goel et al. (1971) contains amino acid data sets which are complete, i.e., the rarer amino acid frequencies have at least been estimated. There are two organisms common to both data sets. Utilizing the Goel group data, I obtain $H_1 = 4.15$ bits for E. coli, which agrees almost exactly with the theoretical value. For M. lysodeikticus, however, I obtain $H_1 = 4.06$ bits, which does not agree as well.

SIGNIFICANCE OF SMITH'S RESULT

Smith's value of 43% (C + G) and King's value of 44.4% (C + G) are in the upper part of the range where all vertebrate DNA is found. It is possible to fix this maximum more accurately; this work is in progress. We discussed the vertebrate base composition range in Chapter 4 in the section on D_1 and D_2. We noted that the base composition of vertebrates is not exactly fixed. It simply lies in the relatively narrow range 42 ± 2% (C + G).

This oscillation about a point rather than exact fixation at that point is indicative of an underlying optimization process. In Chapter 5 on game theory we described in detail the interaction of the game theoretic variables and the nature of the optimization process. The elementary variables being optimized are the familiar elements of variety versus reliability.

If we wish to design a complex living structure, a variety and freedom of choice in the primitive building units, the amino acids, are desired capabilities. The protein entropy, H_1, is a direct measure of this freedom of choice since it is a measure of the degree of equiprobability of the amino acid frequencies.

If all the amino acids in a protein occurred with equal frequency, this protein would have an entropy of $\log_2$ 20 or 4.32 bits. If STOP is considered as a word, then the entropy would be $\log_2$ 21 or 4.39 bits. But under assumptions 1 through 4 the amino acids cannot occur with equal frequency. There is a base composition, however, where this divergence from equiprobability is minimal, and this is the maximum we observe in the plot of $HR(\lambda)$ versus λ. *Given the genetic code and given assumptions 1 through 4, this is the base composition at which the amino acids are more nearly equiprobable than at any other base composition.* When encoding a message, this is the base composition at which there is the greatest freedom of choice of amino acids in protein. Take, for example, the amino acid, phenylalanine. It is coded by UUU and UUC. We could say it is coded mostly by U. In *M. lysodeikticus*, which has a base composition of 71% (C + G), phenylalanine is not one of the more frequently occurring amino acids. For example, it is 2.4% in the data of Sueoka (1961). Even if it were advantageous to the organism to use a much higher percentage of phenylalanine in its proteins, this would not be possible under the direct proportionality assumption simply because there are not enough codons for it in the DNA. At the maximum in H_1 on (21-*S*), an organism has the greatest freedom from such restriction. The preceding discussion assumes that all codons have an equal probability of being used. If we allow some of the codon probabilities to approach zero, it is possible to obtain an exactly equiprobable amino acid frequency distribution. In Chapter 9 we will calculate R values for codes with fewer codons which produce such distributions.

This does not necessarily imply that the amino acid composition of organisms with a base composition in this range will actually have this more nearly equiprobable distribution. It simply means they are not restricted to anything less. However, King and Jukes (1969) have selected an amino acid composition from a sample of vertebrate proteins which they believe is representative. Using this amino acid composition, I obtain $H_1 = 4.20$ bits on $(20\text{-}S)^P$. Sueoka (1961) and Fitch (1964) have measured the amino acid composition of a number of phage and bacteria. For these data, $H_1 = 3.89$ to 4.20 bits on $(20\text{-}S)^P$. Some of these values are listed in Table 14. These limited data need to be extended and reconfirmed, but it appears from the presently available data that the protein of higher organisms in general has higher H_1 on $(20\text{-}S)^P$ or greater message variety than most of the lower organisms.

Therefore, in the history of living systems, at about the time the genetic code was established or shortly afterward, the base composition of prevertebrate organisms became stabilized in the range including 43% $(C + G)$ because, under the conditions expressed at that time in all or some of the assumptions in this calculation, this is the base composition which offers the evolutionary advantage of maximum possible potential message variety in the construction of protein sequences under the present genetic code.

If message variety in the protein were the only desired capability, all vertebrates would have evolved at the exact maximum of H_1 for the protein language. However, diversity is not the only desired capability in a language. The reliability which comes with increased ordering is also desirable. It is the optimization of these two opposing variables that causes the base composition to "oscillate" about the maximum in H_1. This entire process is best illustrated by Fig. 22, the mechanical analogy of a complex system of interacting, opposing variables seeking an optimum (Chapter 5).

Therefore the vertebrates have evolved by "oscillating" about the maximum in H_1 on the protein space because this gives them the maximum leeway in the intricate optimization mechanics which maintains the delicate "balance of life."

The preceding interpretation of the significance of the maximum in $H_1(P)$ is taken from my paper (Gatlin, 1971) reviewing Smith's work. All of the statements made are based on the work completed at that time. Smith confined his studies to one independent DNA variable. King confined his to three. However, there is no reason to stop here. I have now extended studies of this type up to 64 free DNA variables, the number of codons. This work has been done in collaboration with Hans J. Bremmerman and Lucy King of the Mathematics Department at Berkeley, and will be reported in detail elsewhere. In the following section I present an outline of the major results to date.

THE CAPACITY OF THE DNA-TO-PROTEIN CHANNEL

We have defined and studied the information density of DNA and made inferences about the information density of protein. In studying the calculations of Smith and King, we have been studying the transformation from the DNA to the protein information spaces. This can be described within the classical framework of an information processing channel, the DNA-to-protein channel. We defined the general channel in Chapter 4 and Fig. 17. We will now define it in a somewhat more specific manner.

An information processing channel $[\mathscr{A}, \nu, \mathscr{B}]$ is characterized by its input alphabet, $\mathscr{A}$, from the source, its output alphabet, $\mathscr{B}$, and a probabilistic law, ν, connecting the two. We will continue to identify $\mathscr{A}$ as the DNA alphabet at the input and $\mathscr{B}$ as the protein alphabet at the output. In the calculations which follow we will consider the word "STOP" as a part of the protein space. The probabilistic law ν within the channel will consist of the first two assumptions of Smith, the direct proportionality assumption and the genetic code. This is the simplest possible channel we can construct.

When an ergodic source drives a channel, we obtain a definite "rate of transmission" of information whose general functional form is $D_2^{(m)}$. In the self-enclosed section in Chapter 3 we described in detail the subtle differences between our expression and that from classical information theory.

The maximum value which the transmission rate D_2 can take on as we vary the source probability distribution over all possible values is called the ergodic capacity of the channel and represents the upper limit to the rate at which information can be transmitted over the channel without overloading it.

From our basic equation

$$D_2 = H_1 - H_\mathrm{M} \tag{45}$$

it is clear that D_2 takes on its maximum value when $H_M = 0$ and H_1 is maximized as a function of the input symbol probability distribution. Therefore our procedure will be to maximize $H_1(P)$ while allowing the DNA variables at the input to vary freely. The maximum value of $H_1(P)$ so obtained is a biological analog of the classical concept of the channel capacity.

We shall calculate the capacity at three levels or spaces, $S_1(DNA)$, the single letters, $S_2(DNA)$, the doublet sequences and finally $S_3(DNA)$, the triplet sequences or codons themselves. We shall study each space with and without Watson-Crick constraint, making 6 cases in all. We will regard each case as a separate source.

THE SOURCES

Source 1. The simplest possible source is that with $S_1(DNA)$ under Watson-Crick constraint. This is precisely Smith's calculation, which involves one independent variable at the source. Table 15 lists for each source the maximum value of $H_1(P)$ and the base composition of DNA at which it occurs. Also listed are $D_2(DNA)$ and $RD2(DNA)$ at the maximum.

Source 2. If we remove the Watson-Crick constraint from $S_1(DNA)$, we have King's calculation involving three independent variables at the source. King and his co-worker, Glenn Sharrock, performed this calculation using classical gradient optimization techniques which work

TABLE 15

SUMMARY OF SOLUTION VALUES FOR THE DNA-TO-PROTEIN CHANNEL

Source	Number of Variables	%(C + G) at Maximum	$H_1(P)$ (bits)	$D_2(DNA)$ (bits)	R(DNA)	RD2(DNA)
$S_1(DNA)$						
I	1	43.0	4.24	0	.007	0
II	4	42.85	4.25691	0	.011	0
$S_2(DNA)$						
III	10	42.34	4.32855	.123	.070	.878
IV	16	41.19	4.35973	.178	.108	.821
$S_3(DNA)$						
V	32	42.77	4.36903	.417	.217	.962
VI	64	42.72	4.39224	.539	.280	.961
VI[a]	64	40.80	4.39229	.692	.361	.958

[a] Solution and initial guess perturbed randomly up to 100%.

quite well for three variables. However, we now wish to increase the number of variables at the input to 16, the size of $S_2(DNA)$ and finally to 64 for $S_3(DNA)$. More efficient optimization techniques are needed.

OPTIMIZATION TECHNIQUES

To maximize $H_1(P)$ for higher order spaces we have used the highly reliable and effective optimization techniques developed by Bremermann and his co-workers (Bremermann, 1970). Briefly, the method does not require gradients, only evaluations of the function. From an initial estimate, x^o, a random vector, r, is generated, and on the line determined by r and x^o the restriction of the function to this line is approximated by five-point Lagrangian interpolation. The derivative of the interpolation polynomial is a third degree polynomial whose roots, λ_i, are computed by Cardan's formula and the procedure is iterated from the minimum point, $x^o + \lambda_i r$, provided that $F(x^o + \lambda_i r) \leqslant F(x^o)$. To calculate maxima with this program one need only subtract the function from some constant upper bound known to be above the maximum possible value of the function. Bremermann (1970) gives a more complete description of the method, describes its convergence properties, and refers to published applications.

In utilizing this program I place all constraints, except the requirement that the x_i be nonnegative, within the function subroutine and then map from the set of optimization variables to the function variables. This eliminates many bothersome programming difficulties. Thus the number of variables listed for each channel in Table 15 is the actual number of free variables used in the main optimization program although this is not necessarily equal to the number of analytically independent variables in each case.

Source 3. We now consider $S_2(DNA)$ under Watson-Crick constraint. The codon probabilities are calculated under the assumption of a first-order Markov source. The basic variables being optimized are the doublet probabilities. The six Watson-Crick constraints, $p(AC) = p(GT)$, $p(CA) = p(TG)$, $p(AG) = p(CT)$, $p(GA) = p(TC)$, $p(CC) = p(GG)$, $p(AA) = p(TT)$, along with the normalization requirement were placed within the function subroutine. This leaves 10 free variables in the main optimization program.

The 10-variable solution always settles down to the same invariant value after about 200 iterations. Both the initial guess and the final solution were perturbed randomly up to 100%. In every case the function value and the solution set of x_i returned to the same invariant values. This kind of stability is characteristic of a unique solution but does not prove it. The values at the maximum are listed in Table 15.

We note that for Source 3, $S_2(\text{DNA})$ under Watson-Crick constraint, $R(\text{DNA}) = .070$. This theoretical value lies within the range of $R(\text{DNA})$ values which we have calculated from the nearest neighbor data. These are double-stranded data (under Watson-Crick constraint) from a first-order Markov source. Hence our experimental data match the nature of the source we are now considering. The agreement between the theoretical value and the experimental range is rewarding.

Next we constrained the base composition of DNA to specific values and maximized $H_1(\text{P})$ under this additional constraint. Figure 33 shows

FIGURE 33

MAXIMA OF $H_1(\text{P})$.

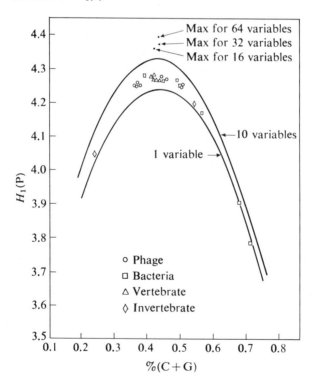

a plot of $H_1(P)$ versus % (C + G) for one variable and 10 variables. The 10-variable curve is the one previously referred to by Smith (1969) as $HM(6, \lambda)^{\text{Max}}$ and the one-variable curve was called $HR(\lambda)$. Our 10-variable curve is similar but not identical to Smith's. We have used different optimization techniques, and also Smith's programming error could have influenced his calculation.

The points on the graph are values of $H_1(P)$ on $(21\text{-}S)^{\text{DNA}}$ (which Smith called H^{Obs}). We see that all of these values lie below the 10-variable maximum curve rather than below the one-variable curve.

Figure 33 also shows the pattern of increasing magnitude of the maximum in $H_1(P)$. As the number of variables at the input of the channel increases, we expect a family of curves of ascending maxima. However, due to limitations on computer time, we calculated these curves explicitly for only two sources.

Source 4. When we remove the Watson-Crick constraint from Source 3, we are left with 16 free optimization variables, the doublet frequencies, which are subject only to the normalization requirement. For this source and all others following, the function values do not reach invariant values as for the 10-variable source. However they do seem to approach asymptotic limits. We began with an equiprobable distribution of frequencies and ran the program for 2000 iterations. For the 16-variable source, the function value has changed only 1.6×10^{-6} bit for the last 500 iterations. We will refer to such solutions as "steady-state" solutions. These values are listed in Table 15.

Source 5. We now consider $S_3(\text{DNA})$ under Watson-Crick constraint. The basic variables are the triplet frequencies. Since each triplet has a Watson-Crick complement and there are no self-complementary triplets, we have 32 optimization variables subject only to the normalization constraint. Our source is a Markov source with a memory of two. Table 15 lists the steady-state solution values after 2000 iterations, starting from an equiprobable triplet frequency distribution. The function value has changed only 3×10^{-4} bit for the last 500 iterations.

Source 6. Finally we remove the Watson-Crick constraint from Source 5, leaving 64 optimization variables subject only to the normalization constraint. This channel bears the closest analogy to the DNA which codes for protein. The steady-state solution is listed in Table 15. The function value has changed only 2.3×10^{-5} bit for the last 500 iterations.

The initial guess and solution after 700 iterations were perturbed randomly up to 100%. Then the program was run for another 700 iterations. This solution is also shown in Table 15.

THE CAPACITY OF THE CHANNEL

Each value of $H_1(P)$ at the maximum is the capacity of the particular channel obtained when the corresponding source drives the channel. These values increase steadily until, in the 64-variable channel, $H_1(P)$ approaches the theoretical limit of $\log_2 21 = 4.3923174$. At this point the amino acids are all equiprobable.

Each value of $H_1(P)$ listed in Table 15 is itself a maximum value of a curve similar to those in Figure 33. Every point on each curve is in turn a maximum value of $H_1(P)$ when the base composition of DNA is constrained to a specific value. Since $H_1(P)$ increases steadily at the maximum as the number of variables at the input of the channel increases, in the highest value of $H_1(P)$ we have a third-level maximum, a maximum of maxima of maxima, another example of the beautiful hierarchical structure of the living system.

Since there are no other constraints in the calculation, the frequency of a particular codon within a given degenerate set is free to vary from 0 to 1/21, provided only that the sum of all the codons for a given amino

TABLE 16

A CODON FREQUENCY DISTRIBUTION FOR $H_1(P) = 4.39227$ BITS					
	U	C	A	G	
	.01034	.00193	.01547	.01412	U
U	.03711	.02337	.03200	.03349	C
	.01531	.00232	.01732	.00492	A
	.00667	.01026	.02541	.04781	G
	.00192	.01567	.00254	.00106	U
C	.01280	.02771	.04514	.01759	C
	.00261	.00331	.00376	.00173	A
	.00835	.00081	.04393	.01028	G
	.01209	.00418	.00891	.00885	U
A	.02900	.02470	.03855	.00069	C
	.00696	.00565	.00906	.00445	A
	.04762	.01300	.03881	.01233	G
	.02114	.00728	.01527	.00160	U
G	.01117	.00738	.03234	.01680	C
	.00843	.00636	.00925	.00873	A
	.00693	.02647	.03854	.02042	G

TABLE 17

A CODON FREQUENCY DISTRIBUTION FOR $H_1(P) = 4.39229$ BITS				
	U	C	A	G	
	.01775	.00450	.02126	.00739	U
U	.03008	.01466	.02592	.04018	C
	.01068	.00155	.01313	.00180	A
	.01991	.01254	.03293	.04716	G
	.00818	.01466	.02130	.01064	U
C	.00169	.01581	.02631	.00145	C
	.00149	.00809	.00551	.00076	A
	.00605	.00977	.04215	.02193	G
	.02762	.00559	.01088	.01012	U
A	.01555	.02226	.03669	.00446	C
	.00435	.00732	.01743	.00391	A
	.04780	.01228	.02988	.00916	G
	.01216	.00060	.00898	.00241	U
G	.00935	.00839	.03863	.00932	C
	.00118	.01164	.01144	.00745	A
	.02487	.02690	.03610	.02807	G

acid is 1/21. Therefore there are an infinite number of codon frequency distributions which are satisfactory solutions. In Tables 16 and 17 I list two typical distributions. The codon frequencies for a given degenerate set are highly unequal. In fact such a highly asymmetric distribution of codon frequencies is necessary to produce the high $R(DNA)$ values we observe for the 64-variable source. Even though the function value changes by only 2×10^{-5} bit, the distributions are quite sensitive to this small change. For example, in Table 16 the codon GUG is the least frequently occurring codon for valine, whereas in Table 17 it is the most frequent. These results are consistent with the calculations of Goel *et al.* (1971), which utilize experimental data to calculate codon frequencies. They conclude that degenerate codons are definitely nonequivalent.

THE OPTIMUM BASE COMPOSITION

The value of the base composition at the capacity of each channel is stable within the range of $42 \pm 2\%$ (C + G). This is the optimum base composition range in which the organism can process information most efficiently. The theoretical range coincides exactly with the experimental base composition range of vertebrates. A base composition of $42 \pm 2\%$ (C + G) corresponds to a range of D_1 values from .010 to .029 bit. Not a single vertebrate value falls outside this range.

FIGURE 34

R(DNA) VERSUS $D_2^{(m)}$ AT THE CAPACITY
OF THE DNA-TO-PROTEIN CHANNEL.

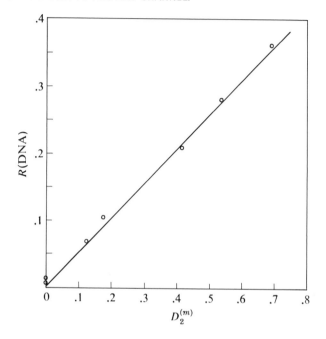

Figure 34 shows a plot of R versus D_2 as we increase the number of source variables. We observe the expected linearity, and the slope is very near the theoretical value of $1/\log a$ for constant D_1.

The relative constancy of D_1 is the result of an averaging process as the codon frequencies vary randomly from 0 to $1/21$ within a given degenerate set. Its particular value is due to the particular structure of the genetic code. We have constructed other codes. They have different base composition ranges at the maximum, but these values are relatively independent of the number of variables at the source and independent of whether the source is under Watson-Crick constraint. The latter independence is expected because the $(C + G)$ sum is the same for the double strand as for each single strand.

THE MAGNITUDE OF R(DNA)

R(DNA) increases from .007 to .361 for the perturbed solution. The magnitude of R(DNA) = .36 is impressive. Let us note the memory of the source as a superscript. We have for the steady-state values of single-stranded DNA: $R^{(0)} = .011$, $R^{(1)} = .108$, $R^{(2)} = .280$. Let us compare these with similar values for human language. From Shannon's (1951) F_N values, which are H_M values in my notation, I calculate that for the English language: $R^{(0)} = .12$, $R^{(1)} = .24$, and $R^{(2)} = .26$. Thus our calculations predict that if the channels we have constructed are realistic models for the living system, in organisms which operate in the vicinity of the channel capacity, the single-stranded DNA which codes for protein has redundancy values of the same order of magnitude as the English language, at least up to the level of a second-order Markov analysis.

These channel calculations are the first to estimate the magnitude of the second-order Markov effect for DNA that utilize no experimental data whatsoever.

We must remember that there is no reason to expect all living systems to operate at *exactly* the capacity of the channel. This would be equivalent to simply maximizing H_1 but, as we have shown, living systems do not seek simple maxima. They seek complex optima, interlocking maxima and minima under constraints, such as the max min of H_1 and the min max of H_M.

We must also remember that we have placed very simple constraints within the channel. In a real living system there must be additional constraints operating. Even so, the redundancy might not be significantly changed because some constraints can increase the redundancy whereas others can lower it, depending on their nature. Generally the constraints or rules of a language increase the redundancy. Usually these rules are of a somewhat arbitrary nature. However there is a type of constraint that can lower the redundancy of a sequence of symbols. This is a symmetry-producing principle such as the Watson-Crick constraint. In every case in Table 15 this constraint lowers the redundancy. Very little is known about symmetry principles in living systems, other than their overt structural expression. Sadler and Smith (1971) have demonstrated a most intriguing symmetry in the genetic structure of an operator.

THE SECOND LAW AND ORDERING

We have been studying, in essence, a model of evolution as an optimization process in a multivariable system. This concept is not new.

But these channel calculations are unique in that they utilize for the first time a "fitness function" (Bremermann *et al.*, 1966) from information theory. The extremum principle is the ancient theme of entropy maximization, but there is a most significant new variation.

The maximization of the entropy $H_1(P)$ at the output of the channel has resulted in the decrease of the entropy $H_M(DNA)$ (at relatively constant H_1) at the input as the number of variables at the source increases.

We have known for some time that the entropy within an isolated system can decrease in localized regions of the system. This does not worry us because we know that the total entropy of the isolated system seeks a maximum in accordance with the second law. But we have never thought of these localized decreases in entropy as being *the direct result* of an entropy maximum principle. We have always regarded them as random fluctuations. This is *not* the situation here. The decrease in the entropy $H_M(DNA)$ is *directly coupled* to the maximization of the entropy $H_1(P)$.

Thus our calculations show that if we place the entropy maximum principle within the constraints of an information processing channel, its power is "harnessed" and put to work in the creation of order. Our third-level entropy maximization at the output (the maximum of maxima of maxima) has increased the redundancy of the sequence of symbols at the input, and this is the sequence which must store the primary hereditary information.

Moreover, random perturbations of the system tend to increase this redundancy further (see Table 15). This is truly order out of disorder.

Furthermore, it is the lack of constraint on the individual codon frequencies within a given degenerate set which gives rise to the high $R(DNA)$ values. To produce a distribution with lower $R(DNA)$ would require severe constraints of a particular nature. Again, this is a higher kind of order resulting from lack of constraint at the proper level of the hierarchy.

The second law of thermodynamics is indeed an order-degrading principle *in itself and without constraint;* but when we place it under the control of the higher laws of information theory, it becomes directly responsible for the production of order of a very important type. This is why life has arisen.

I am among those who believe in the Law of Evolution; I believe that ideal entities evolve, like brute beings, and that religions and governments are raised to higher planes.

The law of evolution has a severe and oppressive countenance and those of limited or fearful mind dread it; but its principles are just, and those who study them become enlightened.

All around me are dwarves who see the giants emerging; and the dwarves croak like frogs.

"The world has returned to savagery. What science and education have created is being destroyed by the new primitives. We are now like the prehistoric cave dwellers. Nothing distinguishes us from them save our machines of destruction and our improved techniques of slaughter."

Thus speak those who measure the world's conscience by their own. They measure the range of all Existence by the tiny span of their individual being. As if the sun did not exist but for their warmth, as if the sea was created for them to wash their feet. —Kahlil Gibran, The Giants

9

SECOND THEOREM EVOLUTION

THE DIRECTION OF EVOLUTION

The theory of Darwin (1909) states that higher forms of life have evolved from lower forms by means of a sifting process called natural selection. Random changes in the genome of the organism are expressed in the structural and, hence, functional features of the organism. Those changes tend to survive or to be selected which give the organism even a slight advantage in a particular environment. We have come to refer to this process as "the survival of the fittest."

However, it has been pointed out that this concept is a tautology. Organisms which are fittest survive and those which survive are fittest. Clearly we need a more definitive explanation of why higher organisms are "fitter."

Manfred Eigen (1971) has developed phenomenological rate equations which describe how information carriers with a "selective value," W_i^o, or growth rate above the average of the ensemble increase. This tends to raise the average until a single species is selected.

Because the solutions to these detailed, non-linear rate equations have a sharp selective behavior quantitatively for information carrier species with the highest values of W_i^o, the conclusion follows that the emergence of self-instructing systems is not the highly improbable, rare event we

191

had previously thought but rather an inevitable occurrence under reasonably existing conditions. This is an exciting new concept.

However, Eigen defines the "information content" of an information carrier simply as $n \log a$. But this is merely the maximum value of the total entropy of *any* sequence of length n. The informational properties of sequences of symbols including their fidelity of replication are clearly a function of more than just their length.

The redundancy of a sequence plays a vital role in the fidelity with which a message can be transmitted over a channel and this channel can be a simple copying process within the context of surrounding molecules. Eigen notes that short sequences tend to survive in his phenomenological box and concludes that the system has a very low "information content." But the point is that these short sequences would tend to have a high *information density* (as our random generator experiments have shown) and hence a higher fidelity of replication, precisely what is needed initially.

If we assume that short sequences (of both DNA and protein) with high R of the high $RD2$ type were the original "survivors" in the phenomenological box *because* they had a higher fidelity of replication, then we have made *a first step in understanding why the fittest are fittest*.

As Eigen points out, value must be defined relative to the context of evolutionary complexity. Once the primitive problem of accurate reproduction has reached some optimal solution, the extremely high S-redundancy is no longer of as much "value" and the system then turns to the solution of the next higher level problem, namely the development of more complex messages with a larger vocabulary.

At this stage the length of the short sequences must begin to increase because this lengthening is a necessary prerequisite for increasing the vocabulary and complexity of a message. In this sense the total information density of a sequence of length n, nI_d, is certainly a fundamental informational parameter of the system. At this stage it is probable that associations of histone-like proteins developed to protect the growing nucleic acid sequences from hydrolysis.

At the stage when the short sequences first begin to grow, we may still reasonably regard them as the output of a random generator under rather light constraint. Under these conditions the R and D_2 values will be monotonically decreasing functions of the length, as shown in Figs. 28 and 29 (Chapter 8). D_1 becomes relatively constant at short lengths because the size of S_1 is small relative to the higher order spaces and D_1 is the first divergence to stabilize.

Thus during the stage when the short sequences first begin to lengthen, D_2 and R will decline in analogy with Figs. 18 and 19 (Chapter 4), which describe the development of vocabulary as the human child learns to read. This process can proceed because once some minimal error detection and correction system based upon pattern recognition has been established, the system can then afford to purchase increased message variety at some reasonable price of increased error probability.

At some point in evolutionary time, the decline of D_2 must have passed through a minimum, because we have observed an increase in D_2 along the $RD2$ vector in the emergence of the vertebrates from lower forms. This is analogous to reversing the direction of the abscissa in Figs. 15 (Chapter 4) and 29 (Chapter 9). The system then begins to climb the D_2 curve in a direction opposite to the "random" direction.

From our basic principles of language evolution, detailed in Chapters 2 and 3, we can predict that D_2 cannot increase without limit. It must level off at some optimum value.

This entire process can be represented as a potential function of the Heitler-London type, except that this time it is D_2 rather than the potential energy which is the fundamental evolutionary variable. This is diagrammed in Fig. 35.

The value of the asymptote of D_2 divides the graph into two domains, as does a vertical line through the minimum. When D_2 is decreasing, we are building vocabulary; when D_2 is increasing, we are building fidelity. Both are of value at the proper stage of evolutionary complexity.

During this entire process the length of the genome has been increasing. Before the minimum in D_2, random generators and the classical concept of the second law of thermodynamics are adequate in explaining the behavior of the self-instructing systems. The nucleic acid sequences would gradually increase in length, approaching $R \leqslant .05$ at about length 100. The protein sequences would, of course, have higher R values for the same length. However, it is the nucleic acid sequences

FIGURE 35

A THEORY OF THE NATURE OF THE CHANGE
IN D_2 WITH EVOLUTIONARY TIME.

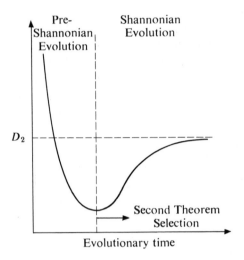

which store the primary information, so we shall focus on these. Thus at the minimum in D_2 we have essentially "random" sequences, i.e., $D_2 \cong 0$ and $D_1 \cong 0$.

However, at the minimum the value of D_2 begins to increase along the $RD2$ vector *while the length of the genome is still increasing*. From this point on, the process is definitely "nonrandom" in the classical sense, and some additional principle over and above the classical concept of the second law of thermodynamics is required to explain the behavior of the self-instructing systems. This higher level principle in the hierarchy of control principles governing living systems is the major theorem of information theory, Shannon's second theorem.

At the minimum of the potential function, programmed control systems begin to emerge. This is made possible by a new type of selection which I call second theorem selection.

The vertical line through the minimum divides the graph into two domains which we might call pre-Shannonian evolution and Shannonian evolution, because it is at the minimum that second theorem selection began to select sequences from among existing ones for the construction of complex programs. It is possible to regard the minimum in the D_2 potential function as a definition of "life," because it was at this point that the $RD2$ vector was reversed. Once the self-instructing systems had reversed the direction set by the classical second law and the random generators and began to climb "time's arrow" in the opposite direction, they were surely "living" systems. The squared Euclidian distance, $R^2 + D_2^2$, is a fundamental distance since it is the length of the $RD2$ vector expressing the evolutionary distance which the vector organisms have traveled along time's arrow.

We may also interpret the reversal in time's arrow as an extension of the classical concept of the second law of thermodynamics. In Chapter 8 we showed that maximization of $H_1(P)$ at the output of the DNA-to-protein channel is directly coupled to the decrease in $H_M(DNA)$ at the input. As is shown in Table 15 (Chapter 8), this is equivalent to the "selection" of DNA sequences with higher R of the high $RD2$ type.

There are many numbers throughout the theory which take on particular physical significance, such as the optimum base composition, the slope of the $RD2$ vector in DNA space, and the minimum recognition length. The value of the asymptote of D_2 is a candidate for another physical constant of unique significance. We may be able to determine this constant as the value which D_2 is approaching asymptotically.

Second theorem selection (or Shannonian selection), which originates at the minimum of the potential function, has no analogy or counterpart in classical evolutionary thought. The remainder of this chapter will be devoted to a discussion of this new type of selection.

We return to our description of evolution as a game. Eigen is keenly aware that game theory holds the "key to any further generalization of evolution theory." He describes a game involving a number of physically acceptable rules which explains why tRNA molecules have the shape of a clover leaf as shown in Fig. 2 (Chapter 1). Einstein was wrong. God *does* play dice with the universe.

We know that the total possible number of genetic messages is "immense" (Elsasser, 1969). Bremermann (1967) estimates this number at $2^{10^{10}}$. Evolution cannot possibly be a search procedure because we cannot possibly search through such a large number of alternatives. This is why evolution is a game. As Eigen points out, the introduction of "rules" can drastically reduce the number of alternative paths of evolution and give it the "definite, effective directional force" (Simpson, 1964) which has impressed so many evolutionary biologists. But now the problem is simply restated unless we can specify precisely what these "rules" of the game are. Therefore I propose that a fundamental "rule" in the evolution of higher organisms from lower forms is the major theorem of information theory, Shannon's second theorem.

We have already stated this principle in general terms in Chapter 4. Let us consider it now from the viewpoint of game theory. Recall that when the max min and the min max are equal the game is said to have a saddle point, as in our example in Fig. 21, the war game (Chapter 5). This does not happen very often.

In a matrix of random numbers the probability of a saddle point is about 1 out of 10. When the max min and the min max are not equal, there is obviously a range of values in between. J. D. Williams (1954) in his delightful book, *The Compleat Strategyst*, states:

> The range between the quantities is the hunting ground where the master player can pick up something more.

This "hunting ground" of game theory where the master player can gain an advantage by more sophisticated techniques of play is an expression of the range of limitation of the generalized second theorem principle. It is only in an intermediate range of the entropy variables that the vertebrate strategy can be implemented. This is why we always place the phrase, "within limits," in the statement of the second theorem principle. As we showed in our discussion of language formulation in Chapter 3, if we carry either variable to the extreme, we "lose the meaning" or the capacity to convey information. Thus we state the second theorem: It is possible *within limited ranges* of the entropy variables to increase the fidelity of a message without loss of potential message variety, provided that the entropy variables change in just the proper way: namely, by increasing D_2 at relatively constant D_1. This is the specific strategy which makes vertebrates master players relative to the lower organisms, and this strategy is implemented within game theoretic limits.

We must now inquire into the detailed mechanism whereby the vertebrates have implemented their strategy. The fundamental underlying mechanism is natural selection, but it is a type of selection which we have not considered previously.

SECOND THEOREM SELECTION

Let us refer again to Fig. 17, the diagram of communications engineering jargon (Chapter 4). Here we have an information processing channel. The source or transmitter is merely any mechanism for generating a sequence of symbols. The encoding of a message in a particular language occurs at the source. The channel is simply any medium over which the message is transmitted and finally received at the output. Conceptually, it is anything one regards as intermediate between transmitter and receiver. We defined the base sequence of DNA as the encoded message at the source of the living channel and the amino acid sequence of proteins as the message which is finally received at the output. This is, of course, in a different language. The channel consists of the entire mechanics of protein synthesis.

All evolutionary thought to date has focused its attention primarily upon the output of this channel, the protein. Under all current thought, natural selection acts because of the sequence of amino acids in proteins. Even the so-called "non-Darwinian" theories of evolution which have arisen recently still focus their attention on the output of the channel, and it is here that they search for the reason why a mutation is selectively neutral, the ultimate reason being that the amino acid in the protein is not critical to the function of the protein.

One can pick up any paper in the evolutionary literature, particularly the more recent ones, and confirm this preoccupation with the output of the channel. For example, I quote from Ohta and Kimura (1971):

> From the point of view of survival probability, the amino acid substitution between a particular pair has a certain average probability of being accepted by natural selection.

Even survival probabilities are conceived in terms of amino acid substitutions in the protein at the output of the channel.

I wish to consider a new type of selection which I shall call second theorem selection because this is the basic principle under which it acts.

Second theorem selection directs our attention for the first time to the input of the channel. I define second theorem selection as natural selection which acts not because of the sequence at the output of the channel but because of the informational efficiency with which this sequence has been encoded at the source under the second theorem principle.

As vertebrates have evolved, they have selected for DNA sequences at the input of the channel with a higher information density of the high $RD2$ type because these sequences have a lower probability of error in the information processing channel, and they achieve this higher measure of fidelity without paying an excessive price for it. This type of selection is made possible because of the extensive degeneracy of the genetic code.

We know that several codons, as many as six in some cases, can code for the same amino acid. This means that for a given protein message at the output of the channel there are a large number of possible DNA sequences at the input of the channel, all of which could code for it. Under current concepts these sequences are all selectively neutral. I quote from King and Jukes (1969):

> Because of the degeneracy of the genetic code, some DNA base-pair changes in structural genes are without effect on protein structure. . . . As far as is known, synonymous mutations are truly neutral with respect to natural selection.

This is not the case with respect to second theorem selection. The different DNA sequences coding for the same amino acid sequence could have significantly different R and D values and hence different probabilities of error in the channel. In order to determine how much influence degenerate codon selection could have on the R- and D-indices, I made the following calculation. I selected arbitrary codes from the accepted dictionary with 20, 30, 40, and 50 codons each. For each given number of codons, I selected three codes, one with high $C + G$, one with intermediate $C + G$, and one with low $C + G$ content by selecting for the third digit of the codon. I kept the quantitative pattern of degeneracy, i.e., the number of codons corresponding to a given amino acid, parallel to the accepted dictionary. I then calculated R and

$RD2$ for encoding a message of equiprobable, independent amino acids for each one of these twelve arbitrarily chosen codes. Table 18 lists D_1, D_2, R, and $RD2$ for single-stranded DNA and Table 19, for double-stranded DNA.

This calculation shows that selective use of degenerate codons can influence dramatically the R and $RD2$ values. There is a difference of .203 between the highest and lowest R values for the single-stranded DNA which is 20.3% of the theoretical range of R. Thus there is adequate variation for second theorem selection to act upon.

The single-strand R values are higher than the double-strand R values. This is expected because the Watson-Crick equivalence relations, $A = T$, $C = G$, tend to symmetrize the transition probability matrix for the double strand, hence reducing D_2, hence reducing R. Therefore it is more efficient to store the genetic information only in one strand of DNA because the single strand can achieve a higher information density.

A fundamental result of information theory is that, the fewer the number of codons used, the lower is the error. In agreement with this, Table 18 shows that, in general, the fewer the number of codons used, the higher is the redundancy, hence the lower the error. It would not be at all surprising if future experimental work shows that, in general, the higher organisms tend to use fewer codons to code for protein. Marshall *et al.* (1965) have found that, *in vitro*, different organisms utilize different codons from the same degenerate set to considerably different extents. On the other hand, the increasing complexity of higher organisms could mean that they use fewer codons to specify amino acids while the remaining codons are used for more and more sophisticated instructional purposes.

Another coding device which can increase R without changing the message variety is simple rearrangement of the codons. In living organisms this would amount to key amino acid rearrangements which result in higher R of the high $RD2$ type as higher organisms evolve. For example, Table 20 lists three sequences. I have calculated R and $RD2$ for long messages composed of repeated units of these three sequences. Sequence II is simply a rearrangement of triplet codons from sequence I. The R and $RD2$ values between I and II are not extremely large; but

TABLE 18[a]

SINGLE-STRAND VALUES OF D_1, D_2, R, AND $RD2$ FOR ENCODING A MESSAGE OF EQUIPROBABLE INDEPENDENT AMINO ACIDS WITH 12 ARBITRARILY CHOSEN CODES

Number of codons		D_1	D_2	R	$RD2$
20	A	.055	.286	.171	.838
	B	.032	.403	.218	.926
	C	.167	.280	.224	.626
30	A	.053	.293	.173	.847
	B	.008	.299	.153	.975
	C	.126	.261	.193	.674
40	A	.022	.298	.160	.932
	B	.031	.153	.092	.830
	C	.124	.172	.148	.582
50	A	.001	.041	.021	.975
	B	.037	.141	.089	.791
	C	.079	.123	.101	.608

[a] Taken from Gatlin (1968).

TABLE 19[a]

DOUBLE-STRAND VALUES OF D_1, D_2, R, AND $RD2$ FOR ENCODING A MESSAGE OF EQUIPROBABLE INDEPENDENT AMINO ACIDS WITH 12 ARBITRARILY CHOSEN CODES

Number of codons		D_1	D_2	R	$RD2$
20	A	.052	.003	.027	.051
	B	.000	.263	.132	1.000
	C	.163	.018	.090	.097
30	A	.048	.003	.025	.051
	B	.007	.111	.059	.939
	C	.126	.031	.078	.198
40	A	.020	.026	.023	.566
	B	.015	.106	.060	.877
	C	.121	.027	.074	.182
50	A	.000	.028	.014	1.000
	B	.027	.079	.053	.746
	C	.061	.042	.051	.410

[a] Taken from Gatlin (1968).

TABLE 20[a]

EXAMPLE SEQUENCES		
Sequence	*R*	*RD2*
I AGT/ATG/CGT/GAC/CAT/CTA ...	.133	.966
II AGT/CGT/CAT/CTA/ATG/GAC ...	.209	.979
III AAA/AAT/TTT/TCC/CCG/GGG ...	.619	.993

[a] Taken from Gatlin (1968).

small differences in error reduction during processing of the genetic information could confer distinct evolutionary advantages.

Thus there are three fundamental coding devices which may increase the reliability of the genetic message: (1) the degenerate codons selected; (2) the number used; and (3) their arrangement. From Table 20, we see that large differences in R and $RD2$ result when entirely different codons and hence amino acids are used, as in sequence III. However, all sequences of Table 20 have the same D_1; they have simply increased D_2 at constant D_1.

The $RD2$ values for the double strand span almost the entire possible range 0 to 1 while the $RD2$ values for the single strand are confined to 30.3% of this range. The specific limits are .05 to 1.0 for double-stranded DNA and .58 to .97 for single-stranded DNA. From Table 5 (Chapter 4) we can calculate that the $RD2$ range of phage is .01 to 1.0, of bacteria, .09 to 1.0 and, of vertebrates, .6 to .8. There is a striking coincidence in these limits.

We must remember that the nearest neighbor data are double-stranded DNA data. Yet our computational model has revealed that single-stranded DNA messages have higher R of the high $RD2$ type and the range of $RD2$ is virtually the same as for vertebrates, whereas double-stranded DNA messages have lower R with the same wide variation in $RD2$ as we observe in lower organisms.

We conclude that vertebrate DNA, although it is double-stranded, has acquired informational properties which mimic those of the single-stranded DNA. There must have been a highly efficient evolutionary process whereby sequences with higher R and $RD2 = .6$ to $.8$ were selected. The "driving force" behind this selection is the "hunting ground" of the second theorem principle, that intermediate range of the entropy variables where the master player can increase the fidelity of the genetic message without excessive penalty.

Thus second theorem selection can distinguish between different DNA base sequences which give rise to the same amino acid sequence in protein. This is a new concept in evolutionary thought. Let us go a step further. It is possible that second theorem selection can distinguish between different DNA sequences which code for slightly different amino acid sequences that are selectively neutral in the Darwinian sense.

It is believed that selectively neutral mutations are fixed by random drift, and it is assumed that the distribution of fixation along the polypeptide axis is random (Kimura, 1962). However, some experimental data do not fit this random distribution (Jukes, 1971). If we impose the concept of second theorem selection as a constraint upon this random distribution model, perhaps we could improve the agreement. This possibility is totally unexplored.

It is conceivable that second theorem selection and Darwinian selection might sometimes conflict. A change in the base sequence of DNA might lower transmission error but produce a protein less fit for its environment, or conversely. In such a case, nature must select for the overall advantage to the organism.

Darwinian selection could also act so as to reduce transmission error in the channel. This would be equivalent to improving the efficiency of a machine by improving its hardware. However, the second law of thermodynamics quickly places an upper bound on improved efficiency by this method. Bremermann (1967) has shown that *E. coli* is a marvel of thermodynamic efficiency and perhaps approaches 100% thermodynamic efficiency. *If improving the hardware of the computer were the only method of improving the efficiency of information processing in the living system, evolution would have reached its apex in E. coli.*

The laws of information theory go far beyond the second law of thermodynamics. We have shown specifically how an entropy maximization at the output of an information processing channel can be directly responsible for a decrease in entropy of the sequence of symbols at the input. The two are directly coupled. The classical view tends to explain localized entropy decreases as random fluctuations rather than specific couplings. In living systems there exists a complex hierarchy of interlocking entropy maxima and minima, and the maximization of one *kind* of entropy can result in the minimization of another. This is an extension of the concept of the second law.

We have shown that invariants on information spaces describe evolution, and we have verified these principles with general language studies. In my opinion, the most exciting possibilities for future work lie in the search for other invariants on information spaces and in the construction of games.

Returning to our game theoretic description, we have shown that the organisms which fall off the $RD2$ vector, "time's arrow," are poor players at the game of life. The higher organisms have acquired and are continuing to acquire the techniques and strategies of a master player. It should be clear that survival *per se* is not the sole criterion of a master player. This is closely analogous to a situation one finds in any large gambling casino. The little old lady at the slot machine is still with us, but she is *not* a master player.

There is a great deal of current speculation by the human species regarding its own ability to survive and continue this evolutionary process at which it has been so successful. In this regard, I often think of Isaac Newton's famous statement that he felt like a child playing with pebbles on the beach while the great ocean of truth lay all undiscovered before him.

If we extend this simile to science and man in general, we are no longer children and we are no longer playing with mere pebbles on the beach. We have been set adrift, in a very small boat, on the vast and stormy sea of discovery. It is up to us to chart our own course. When man looks up, the stars may furnish him a reference frame, but they cannot choose his destination, or steer him clear of the rocks, or protect him from other dangers, because the greatest dangers are within man himself.

We often hear the cry in our age that science is for our destruction. This is half-truth. Science is also for our salvation. Science has simply forced man to face himself, and he is still prostrate upon the deck of his small vessel while the waves of self-incrimination wash over him.

It is time for us to take the helm and accept the responsibility of shaping our own evolution and destiny. We cannot banish Science from our vessel, much like an erring and unwanted servant, and return to our innocent childhood of pebbles on the beach. For Science is not man's servant; *neither is it his enemy*. Science is, in reality, man's most valuable ally in the struggle against "Nature." If we come to identify Science as our opponent in this contest, we are indeed lost; but, if we form a coalition with her, we may yet win the game.

REFERENCES

Abramson, N. (1963). *Information Theory and Coding*, McGraw-Hill, New York.

Apter, M. J., and Wolpert, L. (1963). *J. Theoret. Biol.*, **8**, 244.

Arrighi, F. E., Mandel, M., Bergendahl, J., and Hsu, T. C. (1970). *Biochem. Genet.*, **4**, 367.

Bellamy, A. R., and Joklik, W. K. (1967). *Proc. Nat. Acad. Sci. U.S.A.*, **58**, 1389.

Berger, H., and Yanofsky, C. (1967). *Science*, **156**, 394.

Bergquist, P. L. (1966). *Cold Spring Harbor Symp. Quant. Biol.*, **31**, 435.

Berlekamp, E. A. (1968). *Algebraic Coding Theory*, McGraw-Hill, New York.

Bremermann, H. (1967). In *Progress in Theoretical Biology*, Vol. 1, Academic Press, New York.

Bremermann, H. J. (1970). *Math. BioSci.*, **9**, 1.

Bremermann, H., and Lam, L. S. (1970). *Math. BioSci.*, **8**, 449.

Bremermann, H. J., Rogson, M., and Salaff, S. (1966). In *Fundamental Biological Models*, Edelsak, E. A., Fein, L., Pattee, A. B., and Callahan, A. B., eds., Spartan, Washington, D.C., p. 3.

Brimacombe, R., Trupin, J., Nirenberg, M., Leder, P., Bernfield, M., and Jaouni, T. (1964). *Proc. Nat. Acad. Sci. USA*, **54**, 954.

Britten, R. J., and Davidson, E. H. (1969). *Science*, **165**, 349.

Britten, R. J., and Kohne, D. E. (1967). *Carnegie Inst. Wash. Yearb.*, **66**, 73.

Britten, R. J., and Kohne, D. E. (1968). *Science*, **161**, 529.

Britten, R. J., and Kohne, D. E. (1970). *Scientific Amer.*, **222**, 24.

Burgess, R. R., Travers, A. A., Dunn, J. J., and Bautz, E. K. F. (1969). *Nature*, **221**, 43.

Carbon, J., Squires, C., and Hill, C. W. (1969). *Cold Spring Harbor Symp. Quant. Biol.*, **34**, 505.

Chargaff, E. (1971). *Science*, **172**, 637.

Cheng, T., and Sueoka, N. (1964). *Science*, **143**, 1442.

Cherry, C. (1957). *On Human Communications*, Technology Press of Massachusetts Institute of Technology, Cambridge, Mass.

Christensen, R. A. (1968). Technical Report no. 20, Univ. of California Computer Center, Berkeley, Calif., January.

Clegg, J. B., Weatherall, D. J., and Milner, P. F. (1971). *Nature*, **234**, 337.

Corneo, G., Ginelli, E., and Polli, E. (1970). *J. Mol. Biol.*, **48**, 319.

Crick, F. H. C. (1955). Unpublished note to "RNA Tie Club."

Crick, F. H. C. (1966a). *J. Mol. Biol.*, **19**, 548.

Crick, F. H. C. (1966b). *Cold Spring Harbor Symp. Quant. Biol.*, **31**, 3.

Crick, F. H. C. (1966c). *Scientific Amer.*, **215**, 55.

Crick, F. H. C. (1967). *Of Molecules and Men*, Univ. of Washington Press, Seattle, Wash.

Crick, F. H. C., Griffith, J. S., and Orgel, L. E. (1957). *Proc. Nat. Acad. Sci. U.S.A.*, **43**, 416.

Crick, F. H. C., Barnett, L., Brenner, S., and Watts-Tobin, R. J. (1961). *Nature*, **192**, 1227.

Dancoff, S. M., and Quastler, H. (1953). In *Essays on the Use of Information Theory in Biology*, Univ. of Illinois Press, Urbana, Ill.

Darwin, C. R. (1909). *The Origin of Species*, P. F. Collier and Son, New York.

Davies, J., Gilbert, W., and Gorini, L. (1964). *Proc. Nat. Acad. Sci. U.S.A.*, **57**, 883.

Eigen, M. (1971). *Naturwissenschaften*, **58**, 465.

Elsasser, W. M. (1958). *The Physical Foundations of Biology*, Pergamon Press, New York.

Elsasser, W. M. (1966). *Atom and Organism*, Princeton Univ. Press, Princeton, N.J.

Elsasser, W. M. (1969). *J. Theoret. Biol.*, **25**, 276.

Fitch, W. M. (1964). *Proc. Nat. Acad. Sci. U.S.A.*, **52**, 298.

Garen, A., and Siddiqi, O. (1962). *Proc. Nat. Acad. Sci. U.S.A.*, **48**, 1121.

Gartner, T. K., Orias, E., Lannan, J. E., Beeson, J., and Reid, P. J. (1969). *Proc. Nat. Acad. Sci. U.S.A.*, **62**, 946.

Gatlin, L. L. (1963). *J. Theoret. Biol.*, **5**, 360.

Gatlin, L. L. (1966). *J. Theoret. Biol.*, **10**, 281.

Gatlin, L. L. (1968). *J. Theoret. Biol.*, **18**, 181.

Gatlin, L. L. (1971). *Math. BioSci.*, in press.

Ghosh, H. P., Soll, D., and Khorana, H. G. (1967). *J. Mol. Biol.*, **25**, 175.

Goel, N. S., Rao, G. S., Yčas, M., Bremermann, H. J., and King, L. (1971). *J. Theoret. Biol.*, in press.

Gomatos, P. J., Krug, R. M., and Tamm, I. (1965). *J. Mol. Biol.*, **13**, 802.

Gorini, L., and Kataja, E. (1964). *Proc. Nat. Acad. Sci. U.S.A.*, **51**, 487.

Gorini, L., Jacoby, G. A., and Breckenridge, L. (1966). *Cold Spring Harbor Symp. Quant. Biol.*, **31**, 657.

Hay, J., and Subak-Sharpe, J. H. (1968). *J. Gen. Virol.*, **2**, 469.

Hirsh, D. (1970). *Nature*, **228**, 57.

Hoyer, B. H., and Roberts, R. B. (1967). In *Molecular Genetics*, Academic Press, New York.

Hunter, A. R., and Jackson, R. J. (1970). *Eur. J. Biochem.*, **15**, 381.

Jacob, F. (1966). *Science*, **152**, 1470.

Jacob, F., and Monod, J. (1961). *J. Mol. Biol.*, **3**, 318.

Jacob, F., Ullman, A., and Monod, J. (1964). *Compt. Rend.*, **258**, 3125.

Jacobson, K. B. (1971). *Progr. Nucleic Acid Res. Mol. Biol.*, **11**, 461.

Jones, D. D. and Dayhoff, M. O. (1972). *Biophysical Society Abstracts*, **12**, 162a, abstr. no. SaPM-C11.

Jones, K. W. (1970). *Nature*, **225**, 912.

Josse, J., Kaiser, A. D., and Kornberg, A. (1961). *J. Biol. Chem.*, **236**, 864.

Jukes, T. H. (1966). *Molecules and Evolution*, Columbia Univ. Press, New York.

Jukes, T. H. (1971). *Proc. 6th Berkeley Symp. Math. Stat. Prob.*, in press.

Jukes, T. H., and Gatlin, L. (1970). *Prog. Nucleic Acid Res. Mol. Biol.*, **11**, 303.

Khinchin, A. I. (1957). *Mathematical Foundations of Information Theory*, Dover, New York.

Kimura, M. (1968). *Nature*, **217**, 624.

King, J. L. (1971). *Proc. 6th Berkeley Symp. Math. Stat. Prob.*, in press.

King, J. L., and Jukes, T. H. (1969). *Science*, **164**, 788.

Koestler, A. (1967). *The Ghost in the Machine*, Macmillan, New York.

Kohne, D. (1970). *Quart. Rev. Biophys.*, **33**, 327.

Kornberg, A., Bertsch, L. L., Jackson, J. F., and Khorana, H. G. (1964). *Proc. Nat. Acad. Sci. U.S.A.*, **51**, 315.

Krakow, J. S., Daley, K., and Karstadt, M. (1969). *Proc. Nat. Acad. Sci. U.S.A.* **63**, 423.

Krzywicki, A., and Slonimski, P. P. (1966). *Compt. Rend.*, **262**, 515.

Kurland, C. G. (1970). *Science*, **169**, 1171.

Lodish, H. F., and Robertson, H. D. (1969). *Cold Spring Harbor Symp. Quant. Biol.*, **34**, 655.

McGeoch, D. J. (1970). Ph.D. dissertation, University of Glasgow, Scotland.

McGeoch, D. J., Crawford, L. V., and Follett, E.A.C. (1970). *J. Gen. Virol.*, **6**, 33.

Malkin, L. I., and Rich, A. (1967). *J. Mol. Biol.*, **26**, 329.

Marshall, R. E., Caskey, C. T., and Nirenberg, M. (1965). *Science*, **155**, 820.

Morrison, J. M., Keir, H. M., Subak-Sharpe, J. H., and Crawford, L. V. (1967). *J. Gen. Virol.*, **1**, 101.

Muench, K. H., and Berg, P. (1966). *Biochemistry*, **5**, 970.

Neyman, J., and Scott, E. L. (1957). In *The Universe*, Simon and Schuster, New York.

Nichols, J. L. (1970). *Nature*, **225**, 147.

Nirenberg, M. W., and Matthaei, H. (1961). *Proc. Nat. Acad. Sci. U.S.A.*, **47**, 1588.

Nishimura, S., Jones, D. S., Ohtsuka, E., Hayatsu, H., Jacob, T. M., and Khorana, H. G. (1965). *J. Mol. Biol.*, **13**, 283.

Nomura, M., Mizushima, S., Ozaki, M., Traub, P., and Lowry, C. V. (1969). *Cold Spring Harbor Symp. Quant. Biol.*, **34**, 49.

Ohta, T., and Kimura, M. (1971). *Nature*, in press.

Ozaki, M., Mizushima, S., and Nomura, M. (1969). *Nature*, **222**, 333.

Parzen, E. (1962). *Stochastic Processes*, Holden-Day, Inc., San Francisco, Calif., p. 191.

Pattee, H. H. (1967). *Towards a Theoretical Biology*, Waddington, C. H., ed., Aldine, Chicago, Ill.

Polanyi, M. (1966). *The Tacit Dimension*, Doubleday, New York.

Polanyi, M. (1967). *Chem. Eng. News*, **45**, 54.

Quastler, H. (1964). *The Emergence of Biological Organization*, Yale Univ. Press, New Haven, Conn.

Rechler, M. M., and Martin, R. G. (1970). *Nature*, **226**, 908.

Reichert, T. A., and Wong, A. K. C. (1971). *J. Mol. Evolution*, **1**, 97.

Roberts, J. W. (1969). *Nature*, **224**, 1168.

Russell, B., and Whitehead, A. N. (1910). *Principia Mathematica*, Cambridge Univ. Press, New York.

Sadler, J. R., and Smith, T. F. (1971). *J. Mol. Biol.*, **62**, 139.

Scholtissek, C., and Rott, R. (1969). *J. Gen. Virol.*, **4**, 565.

Seuss, Dr. (1955). *On Beyond Zebra*, Random House, New York.

Shannon, C. E. (1949). In C. E. Shannon and W. Weaver, *The Mathematical Theory of Communication*, Univ. of Illinois Press, Urbana, Ill.

Shannon, C. E. (1951). *Bell System Tech. J.*, January, p. 50.

Simpson, G. G. (1964). *This View of Life*, Harcourt, Brace and World, New York, p. 164.

Sivolap, Y. M., and Bonner, J. (1971). *Proc. Nat. Acad. Sci. U.S.A.*, **68**, 387.

Smith, T. F. (1969). *Math. BioSci.*, **4**, 179.

Soll, D., Ohtsuka, E., Jones, D. S., Lohrmann, R., Hayatsu, H., Nishimura, S., and Khorana, H. G. (1965). *Proc. Nat. Acad. Sci. U.S.A.*, **54**, 1378.

Southern, E. M. (1970). *Nature*, **227**, 794.

Southern, E. M. (1971). *Nature New Biol.*, **232**, 82.

Steitz, J. A. (1969). *Nature*, **224**, 957.

Stent, G. S. (1963). *Molecular Biology of Bacterial Viruses*, W. H. Freeman and Co., San Francisco, Calif.

Subak-Sharpe, J. H. (1969a). *Can. Cancer Conference*, **8**, 242.

Subak-Sharpe, J. H. (1969b). In *Handbook of Molecular Cytology*, Lima-de-Faria, A., ed., North-Holland Publishing Co., Amsterdam.

Subak-Sharpe, J. H., Burk, R. R., Crawford, L. V., Morrison, J. M., Hay, J., and Keir, H. M. (1966). *Cold Spring Harbor Symp. Quant. Biol.*, **31**, 737.

Sueoka, N. (1960). *J. Mol. Biol.*, **3**, 31.

Sueoka, N. (1961). *Proc. Nat. Acad. Sci. U.S.A.*, **47**, 1141.

Sueoka, N. (1964). In *The Bacteria*, Vol. V, Gunsalus, I. C., and Stanier, R. Y., eds., Academic Press, New York, p. 421.

Sueoka, N. (1965). In *Evolving Genes and Proteins*, Bryson, V., and Vogel, H. J., eds., Academic Press, New York, p. 480.

Sutton, W. D., and McCallum, M. (1971). *Nature New Biol.*, **232**, 83.

Swartz, M. N., Trautner, T. A., and Kornberg, A. (1962). *J. Biol. Chem.*, **237**, 1961.

Takanami, M., Yan, T., and Jukes, T. H. (1965). *J. Mol. Biol.*, **12**, 761.

Temin, H. M., and Mizutani, S. (1970). *Nature*, **226**, 1211.

Thomas, C. A., Jr. (1966). *Prog. Nucleic Acid Res. Mol. Biol.*, **5**, 315.

Traub, P., and Nomura, M. (1968). *Science*, **160**, 198.

Von Neumann, J. (1955). *Mathematical Foundations of Quantum Mechanics*, Princeton Univ. Press, Princeton, N.J.

Von Neumann, J., and Morgenstern, O. (1947). *Theory of Games and Economic Behavior*, Princeton Univ. Press, Princeton, N.J.

Weaver, W. (1949). In C. E. Shannon and W. Weaver, *The Mathematical Theory of Communication*, Univ. of Illinois Press, Urbana, Ill.

Webster, R. E., and Zinder, N. D. (1969). *J. Mol. Biol.*, **42**, 425.

Weiss, G. B. (1970). *Math. BioSci.*, **8**, 291.

Weiss, S. B., and Nakamoto, T. (1961). *Proc. Nat. Acad. Sci. U.S.A.*, **47**, 1400.

Wigner, E. P. (1967). *Symmetries and Reflections*, Indiana Univ. Press, Bloomington, Ind.

Williams, J. D. (1954). *The Compleat Strategyst*, McGraw-Hill, New York.

Yčas, M. (1958). In *Symposium on Information Theory in Biology*, Yockey, H. P., Platzmann, R. L., and Quastler, H., eds., Pergamon Press, New York, p. 70.

Yčas, M. (1969). *The Biological Code*, North-Holland Publishing Co., Amsterdam.

INDEX